DACE Price Booklet

Cost information for estimation and comparison

Edition 35, November 2021

dace
cost and value

DACE Cost and Value - The Dutch Network and Knowledge Center for Cost Engineering and Value Management
Ambachtsstraat 15, 3861 RH Nijkerk
PO box 1058, 3860 BB Nijkerk
The Netherlands
Tel.: +31 ((0)33) 247 34 55
e-mail: info@dace.nl
internet: www.dace.nl
www.dacepricebooklet.com

ACTIVATION CODE 325C85025A

Please navigate to 'register and activate' on www.dacepricebooklet.com, follow the instructions and provide the code. You will get the online booklet on your profile.

COLOPHON

DACE Price Booklet
35th edition, November 2021
A publication of Bouwkosten.nl B.V., The Hague, The Netherlands

Authors
DACE Cost and Value - The Dutch Network and Knowledge Center for Cost Engineering and Value Management

Address
P.O. Box 1058, 3860 BB Nijkerk

Publisher
Bouwkosten.nl BV

Service Desk
Telephone +31 (0)85 01 30 471
E-mail: bouwkosten@igg.nl

General conditions
All offers, quotations and agreements of Bouwkosten.nl BV are subject to the conditions filed with the Chamber of Commerce in The Hague, The Netherlands.

INTRODUCTION

We take pride in presenting the 35th edition of the DACE Price Booklet. It has been edited by DACE in cooperation with Bouwkosten.nl BV, The Hague, The Netherlands. DACE cost information can be accessed online on www.dacepricebooklet.com. This requires a one-time registration using the unique activation code on page 2.
Bouwkosten.nl BV, a part of IGG bouweconomie BV, takes the responsibility for production, marketing and distribution, while DACE is responsible for the content. Cost information is collected by the Special Interest Group CEPI (Cost Engineering Process Industry). Just as in former editions, the authors have compiled this information with the utmost care. Budget prices have been established from judgement of the actual costs of recently completed projects. The useful value of this document might possibly increase if users' wishes can be taken into account. In case you have any remarks or suggestions for the lay-out, completeness, ease of use or otherwise, we kindly invite you to inform Bouwkosten.nl BV (bouwkosten@igg.nl). For questions regarding the content of the DACE Price Booklet, please contact the DACE office (info@dace.nl).

DACE Cost and Value: - The Dutch Network and Knowledge Center for Cost Engineering and Value Management.
DACE is an organization in which more than 100 companies participate. These companies operate in the Dutch process industry and infrastructure industry. Cost engineers and value managers from these companies participate in the DACE networking meetings, the Special Interest Groups, and in other events. All activities focus on building an effective and efficient network of cost engineers and value managers. Visit www.dace.nl for more information.

Nijkerk, The Netherlands
November 2021

DACE Cost and Value
The Dutch Network and Knowledge Center for Cost Engineering and Value Management.

DIRECTIONS FOR USERS

For what purpose can the DACE Price Booklet be used?

Provided costs (prices) are applicable within the process industry; intended to be used for budget reasons, or for comparing alternative project implementation methods. Costs are not suitable for definitive estimates or benchmarking purposes. If for example, large quantities are used then special discounts are often possible on the listed prices.

How to use this DACE Price Booklet?

All prices are:

- Applicable within the process industry;
- Intended to be used for budget reasons, or for comparing alternative project implementation methods;
- Not suitable for definitive estimates or benchmarking purposes;
- Based on West European construction method, cost level based on The Netherlands;
- Budget prices, exclusively for the use of cost estimation and comparison of costs for alternative methods of execution based on wide accuracy range, and not suitable for definitive estimates;
- Expressed in Euros and exclusive of VAT;
- Based on the price level 4nd quarter 2021, unless stated otherwise, the full impact of the coronavirus (COVID-19) has not been taken into consideration;
- Exclusive of usual trade discounts, granted to industrial users, unless stated otherwise;
- Exclusive of special discounts while ordering larger quantities;
- Generally depending on various details such as accepted risk profile (ALARP), shape, quality or type, whilst for the sake of larger quantities special discounts are usually possible;
- Related to activities, carried out by contractors or installers, assuming a process industry HSE profile;
- For activities performed by contractors, based on a normal market situation;
- Assuming a normal market situation.

Dimensions are indicated in millimeters, any deviations are clearly stated.
Generally the SI Unity system is used.

For which region and which year are the pricing levels the basis in the DACE Price Booklet?

These budget costs are for West European circumstances and based on the 4Q2021 pricing level.

Could DACE provide additional details or clarifications of existing pricing tables?

If you see opportunities to improve specific tables, or notice discrepancies, then please provide us with feedback. Upon receipt we shall evaluate your input, and will do our best endeavor to incorporate these possible improvements in the next DACE Price Booklet release. Please contact us at info@dace.nl

Is it possible to initiate and add new cost tables?

Yes, we recommend new ideas and opportunities. Please submit an email to info@dace.nl, including sufficient back-up, and DACE will contact you within 2 weeks.

Some tables do not include costs, e.g. B4002, E0001, F0001. Could you explain the objective of these tables?

The content of these general sections is to clarify the subsequent cost tables.

We see some discrepancies in costs/prices of previous DACE Price Booklet editions. Is it possible to update the older Price Booklet editions or provide clarifications of these previous editions?

Unfortunately we cannot and will not re-issue previous Price Booklet editions. The intention is to continuously improve the content of this booklet and to incorporate learning's within new released editions.

DIRECTIONS FOR USERS

Could you describe the cost-data collection process including DACE responsibility for new DACE Price Booklet editions?
DACE consists of a large group of volunteers – originating from owners/operating companies, EPC engineering companies and construction contractors within the process industry - who are providing cost input. DACE acts as a facilitator, and is performing some high level process checks on the information received.
Note:
Neither DACE, nor the publisher, accept any form of liability for damage resulting from the use of information provided by means of the "DACE Price Booklet". Cost should not to be used in isolation and the publisher strongly recommends independent verification of key information critical to final use.

What is the background of the indexes outside The Netherlands listed in table H5002?
These indexes are derived from outside organizations such as "ChemieTechnik (Germany)" and "Chemical Engineering (USA)".

What is the background of the Dutch indexes listed in table H5003?
These indexes are derived from the CBS (refer to www.cbs.nl Statistics Netherlands for more details), except series "30" and "32" (refer to www.bouwkosten.nl – for additional details) and "60" (Webci composite index) is a combined index for a process installation built in The Netherlands.

I bought the booklet, where can I get my online subscription?
Please navigate to "register and activate" on www.dacepricebooklet.com and provide the activation code you've find in this book, you will get the online booklet on your profile.

TABLE OF CONTENTS

TABLE OF CONTENTS

TABLE OF CONTENTS

TABLE OF CONTENTS

TABLE OF CONTENTS

TABLE OF CONTENTS

TABLE OF CONTENTS

TABLE OF CONTENTS

A1 Static process equipment

A1001 Atmospheric cylindrical storage tanks

Horizontal or vertical tanks.
PVC polyester
PVC polyester with glass fibre reinforced, approximately 70% glass.
Design with spherical head.
Polyester
Full polyester.
Including:
- Manhole with cover;
- Several nozzles.

Variation: Different L/D relations do have an impact on price while having the same volume.

Content in m^3	Cost each (x 1,000)	
	PVC-polyester	Polyester
1	14.-	14.-
5	20.-	18.-
10	22.-	21.-
20	28.-	26.-
40	35.-	30.-
60	41.-	36.-
80	53.-	41.-
100	67.-	50.-
200	99.-	91.-
300	127.-	107.-
400	156.-	139.-
500	184.-	163.-
600	215.-	195.-
800	277.-	257.-
1,000	320.-	289.-

A1 Static process equipment

A1002 Vertical storage tanks

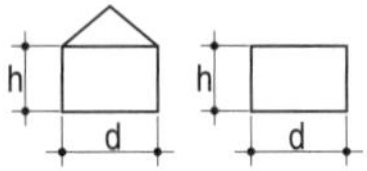

Material: Carbon steel.
Type: Cone roof or floating roof.
Including:
- 1 Outside cage ladder;
- 2 Manholes;
- Several nozzles;
- Hand railing;
- Assembly on site.

Excluding:
- Internals;
- Civil work, painting and insulation.

Design	Volume in m^3	Height (h) in m	Diameter (d) in m	Total surface area in m^2	Weight (x 1,000 kg)	Cost each (x 1,000)
Cone roof	50	5	3.6	77	3.65	45.-
	100	6	4.6	120	5.65	55.-
	150	6	5.6	155	7.3	60.-
	200	7.5	5.8	190	9	71.-
	250	7.5	6.5	220	10.4	80.-
	300	7.5	7.1	247	11.7	87.-
	350	8	7.5	277	13.1	93.-
	400	8	8.0	302	14.25	103.-
	500	9	8.5	355	16.7	112.-
	1,000	9.15	12.2	585	38	201.-
	1,800	10.7	14.6	825	54	269.-
	6,000	12.9	24.4	1,923	137	595.-
	13,500	14.7	34.2	3,415	290	1,105.-
	17,500	14.7	39.1	4,205	380	1,375.-
	25,000	16.5	44.0	5,319	520	1,830.-
	31,000	16.5	48.8	6,267	640	2,300.-
	53,500	18.3	61.3	9,422	1,070	3,865.-
Floating roof	1,000	9.15	12.2	585	50	397.-
	1,800	10	14.6	825	67	443.-
	6,000	12.9	24.4	1,923	158	818.-
	13,500	14.7	34.2	3,415	290	1,275.-
	17,500	14.7	39.1	4,205	385	1,595.-
	25,000	16.5	44.0	5,319	530	2,020.-
	31,000	16.5	48.8	6,267	650	2,440.-
	53,500	18.3	61.3	9,422	1,050	3,780.-
Sphere	610		10.5		100	1,030.-
	3,600		19.0		430	2,905.-

A PROCESS EQUIPMENT

A1 Static process equipment

A1003 Vertical silo's aluminium or AISI 304

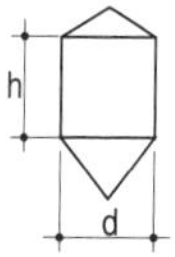

Material: Aluminium or AISI 304.
Type: Conical roof or flat roof, conical bottom, cone angle 90°.
Design pressure: +0.20/-0.02 bar.
Design temperature: 60°C.
Bulk density: 500 kg/m³.
Wall thickness: Average wall thickness of the cylindrical shell.
Price and weight are including:

- Skirt;
- Manhole;
- Several nozzles;
- Cage ladder in carbon steel;
- Handrail in carbon steel;
- Platform in carbon steel.

Excluding:

- Transport and installation;
- Foundation;
- Structural steel.

Material	Content in m^3	Height (h) in m	Diameter (d) in m	Average plate thickness in mm	Weight in kg	Cost each (x 1,000)
Aluminium	10	6.6	1.4	4	690	23.-
				5	770	23.-
	20	8.0	1.8	5	1,170	24.-
				6	1,300	26.-
	40	9.0	2.4	5	1,800	27.-
				6	2,000	30.-
	60	8.6	3.0	5	2,300	36.-
				6	2,600	38.-
	100	10.0	3.6	6	3,700	49.-
				7	4,100	53.-
	200	14.6	4.2	7	6,600	74.-
				8	7,200	81.-
	300	14.0	5.3	8	9,300	99.-
				10	11,000	116.-
	400	16.0	5.7	9	11,500	123.-
				10	13,500	146.-
	500	18.0	6.0	8	13,800	150.-
				10	16,100	173.-
AISI 304	10	6.6	1.4	4	1,860	33.-
	20	8.0	1.8	4	2,840	47.-
	40	9.0	2.4	5	5,030	64.-
	60	8.6	3.0	7	8,540	86.-
	100	10.0	3.6	8	12,980	117.-
	200	14.6	4.2	10	24,460	169.-
	300	14.0	5.3	13	40,450	253.-
	400	16.0	5.7	14	51,740	320.-
	500	18.0	6.0	15	63,700	391.-

A1004 Vertical silo's in plastic material

A

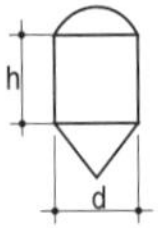

Material: Glass reinforced plastic (polyester).
Type: Spherical roof, conical bottom, cone angle 60°.
Design pressure: +0.20/-0.02 bar.
Design temperature: 60°C.
Bulk density: 500 kg/m³.
Including:
- Skirt;
- Manhole;
- Several nozzles;
- Cage ladder;
- Handrail;
- Platform.

Excluding:
- Transport and installation;
- Foundation;
- Structural steel.

Content in m^3	Height (h) in m	Diameter (d) in m	Cost each (x 1,000)
10	4.9	1.9	9.-
20	5.9	2.4	10.-
40	8.1	2.4	11.60
60	11.6	2.8	19.60
100	12.3	3.5	21.60
200	18.9	4.0	64.80

A1 Static process equipment

A1005 Cylindrical vessels with hemispherical head

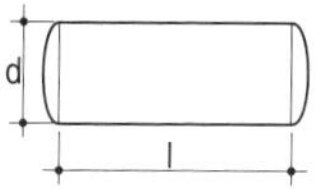

Horizontal or vertical vessels.
Including:
- Supports;
- X-ray;
- Inspection;
- Blasting in- and outside;
- 1 Manhole;
- Maximum 9 nozzles;
- Painting outside;
- Reinforcement rings if necessary.

Excluding:
- Heat treatment.

Variation: Different L/D relations do have an impact on price while having the same volume.

Content in m^3	Length (l) in m	Diameter (d) in m	Total surface area in m^2	Average plate thickness in mm	Weight in kg	Cost each (x 1,000)	
						Carbon steel	AISI 304
1	1.1	1.0	5.3	3	255	12.45	17.35
				5	405	13.35	18.35
				8	615	14.20	21.20
				10	745	14.20	
				13	940	16.-	
				16	1,130	16.90	
5	1.7	1.8	15.3	3	670	19.55	27.-
				5	1,045	24.-	33.75
				8	1,575	27.55	40.50
				10	1,915	30.20	
				13	2,405	31.10	
				16	2,910	33.75	
10	2.1	2.3	25.0	3	1,010	23.10	33.75
				5	1,570	24.90	36.65
				8	2,370	26.65	43.40
				10	2,900	28.45	
				13	3,700	30.20	
				16	4,485	34.65	
20	4.5	2.3	38.0	3	1,515	28.45	45.35
				5	2,365	32.90	51.15
				8	3,630	36.45	61.75
				10	4,465	39.10	
				13	5,685	41.75	
				16	6,930	44.45	
40	5.2	3.0	67.0	3	2,235	34.65	54.-
				5	3,550	39.10	64.65
				8	5,475	45.30	82.95
				10	6,755	48.85	
				13	8,660	55.10	
				16	10,510	59.55	
60	8.0	3.0	91.0	5	4,725	46.20	78.15
				8	7,335	56.85	102.-
				10	9,045	63.10	117.-
				13	11,570	71.10	
				16	14,255	79.10	
80	10.9	3.0	118.0	8	9,215	65.75	122.-
				10	11,345	74.65	145.-
				13	14,765	85.30	
				16	18,195	94.20	
100	13.7	3.0	145.0	8	10,985	83.55	153.-
				10	13,720	94.20	180.-
				13	17,855	107.-	214.-
				16	22,000	118.-	

A PROCESS EQUIPMENT

A1 Static process equipment

A1006 Columns

A

Including:
- Blasting in- and outside, X-ray, Inspection, Painting outside one coat, skirt, nozzles, manholes.

Excluding:
- Platforms, cage ladders, transport, fillings, internals.

Average wall thickness in mm	Average length between the tangentlines (l) in m	Diameter (d) in m	Weight (x 1,000 kg)	Cost each (x 1,000)		
				Carbon steel	AISI 304	AISI 316 L
8	5	0.5	1.1	16.-	87.-	117.-
		1.0	2.5	27.-	143.-	183.-
		1.5	4.2	42.-	194.-	241.-
		2.0	5.7	51.-	222.-	286.-
		2.5	7.6	62.-	265.-	335.-
		3.0	10.2	78.-	367.-	412.-
10	10	0.5	2.3	25.-	146.-	175.-
		1.0	4.6	43.-	214.-	255.-
		1.5	7.2	60.-	284.-	324.-
		2.0	9.6	75.-	320.-	377.-
		2.5	12.4	92.-	377.-	433.-
		3.0	15.9	110.-	425.-	498.-
		3.5	19.2	128.-	449.-	546.-
		4.0	22.9	145.-	486.-	603.-
12	15	0.5	3.8	37.-	208.-	240.-
		1.0	7.2	60.-	305.-	355.-
		1.5	10.9	82.-	351.-	402.-
		2.0	14.4	103.-	380.-	470.-
		2.5	18.4	123.-	435.-	536.-
		3.0	23.1	147.-	488.-	603.-
		3.5	27.8	168.-	535.-	670.-
		4.0	32.7	191.-	582.-	738.-
14	20	1.0	10.5	80.-	348.-	393.-
		1.5	15.6	108.-	403.-	488.-
		2.0	20.6	133.-	488.-	575.-
		2.5	25.9	159.-	619.-	738.-
		3.0	32.1	188.-	573.-	728.-
		3.5	38.3	215.-	629.-	805.-
		4.0	44.7	241.-	666.-	871.-
16	25	1.0	14.3	102.-	394.-	470.-
		1.5	21.3	138.-	469.-	585.-
		2.0	28	169.-	526.-	670.-
		2.5	34.7	200.-	573.-	757.-
		3.0	42.7	233.-	704.-	862.-
		3.5	50.6	265.-	741.-	929.-
		4.0	59	298.-	826.-	1,015.-
18	30	1.0	18.9	126.-	516.-	593.-
		1.5	27.9	169.-	535.-	670.-
		2.0	36.4	206.-	619.-	776.-
		2.5	45.2	244.-	704.-	871.-
		3.0	54.9	282.-	789.-	977.-
		3.5	65.1	321.-	873.-	1,065.-
		4.0	75.7	361.-	949.-	1,160.-
20	35	1.5	35.3	203.-	601.-	766.-
		2.0	45.9	245.-	694.-	881.-
		2.5	57.2	291.-	798.-	996.-
		3.0	69.1	337.-	911.-	1,100.-
		3.5	81.7	382.-	1,015.-	1,205.-
		4.0	94.6	426.-	1,115.-	1,305.-
22	40	1.5	40.3	222.-	666.-	823.-
		2.0	52.2	272.-	769.-	948.-
		2.5	64.7	320.-	882.-	1,065.-
		3.0	77.6	367.-	986.-	1,170.-
		3.5	91.5	417.-	1,100.-	1,285.-
		4.0	105.7	465.-	1,230.-	1,430.-

A1 Static process equipment

A1007 Internals

Description	Diameter in m	Cost per m^2	
		AISI 410 S	AISI 316
Valve trays	1	879.-	650.-
	2	833.-	586.-
	3	798.-	525.-
	4	746.-	490.-
Sieve trays	1	833.-	560.-
	2	798.-	506.-
	3	736.-	435.-
	4	710.-	409.-

A1008 Filling

Description	Material	Diameter in mm	Contents in m^3	Cost per m^3
Pall ring	Carbon Steel	16	1	3,600.-
			10	3,140.-
		25	1	1,530.-
			10	1,050.-
		38	1	1,050.-
			10	610.-
		50	1	870.-
			10	480.-
	AISI 304	16	1	4,490.-
			10	3,920.-
		25	1	3,360.-
			10	2,810.-
		38	1	2,720.-
			10	2,110.-
		50	1	2,350.-
			10	1,820.-
Ceramic balls		6	1	3,450.-
			5	2,600.-
		13	1	2,450.-
			5	1,990.-
		19	1	1,890.-
			5	1,380.-
		25	1	1,790.-
			5	1,280.-

A1 Static process equipment

A1009 Carbon steel heat exchangers

A

Pipe bundle and shell of carbon steel, fixed tube sheet, without expansion bellows, Ø 20 mm.
Design: 'U' tube bundle 10 - 15% cost increase and 5 - 12% weight increase.

Heated surface in m^2	Weight in kg	Cost each (x 1,000)
2	200	14.20
4	300	17.75
6	400	19.55
8	450	20.45
10	500	20.45
15	800	24.-
20	900	25.75
30	1,300	27.55
40	1,600	28.45
50	1,900	32.-
70	2,600	33.75
100	3,600	40.90
200	6,500	55.10
300	10,000	67.55
500	15,000	99.50
700	21,000	135.-
1,000	29,000	161.-

A1010 Stainless steel heat exchangers

Pipe bundle and shell of stainless steel, fixed tube sheet, without expansion bellows, Ø 20 mm.
Variation: 'U' tube bundle 10 - 15% cost increase and 5 - 12% weight increase.

Heated surface in m^2	Mass in kg	Cost each (x 1,000)	
		AISI 304	AISI 316L
2	150	21.-	22.-
4	250	24.-	26.-
6	300	27.-	29.-
8	400	29.-	30.-
10	450	29.-	31.-
15	600	33.-	35.-
20	700	36.-	38.-
30	1,000	39.-	41.-
40	1,200	42.-	45.-
50	1,500	48.-	51.-
70	1,900	52.-	59.-
100	2,500	66.-	73.-
200	4,700	87.-	99.-
300	6,600	113.-	124.-
500	11,000	172.-	190.-
700	15,000	234.-	257.-
1,000	20,000	292.-	318.-

A1 Static process equipment

A1011 Combination carbon steel and stainless steel heat exchanger

Type: Shell of carbon steel, rest stainless steel, fixed tube sheet, without expansion bellows, Ø 20 mm.
Design: 'U' tube bundle 10 - 15% cost increase and 5 - 12% weight increase.

Heated surface in m^2	Cost each (x 1,000)		
	Material pipe bundles		
	AISI 304	AISI 316L	Hastelloy 'B'
2	17.-	18.-	40.-
4	20.-	21.-	46.-
6	22.-	24.-	52.-
8	23.-	24.-	56.-
10	24.-	25.-	61.-
15	25.-	27.-	77.-
20	28.-	29.-	88.-
30	32.-	33.-	118.-
40	33.-	35.-	137.-
50	37.-	39.-	165.-
70	40.-	43.-	200.-
100	56.-	59.-	242.-
200	80.-	84.-	378.-
300	105.-	110.-	565.-
500	149.-	158.-	942.-
700	195.-	208.-	
1,000	263.-	278.-	

A1012 Plate heat exchangers

Type: Plate heat exchanger of one material, including plates, packing and compression bolts.
Working pressure up to 1 MPa, temperature up to 250 °C.
Plate area is the sum of all plates per exchanger.

Plate surface in m^2	Cost each (x 1,000)	
	AISI 316	Titanium
40	23.-	80.-
50	24.-	90.-
60	26.-	103.-
70	28.-	106.-
80	31.-	112.-
100	35.-	117.-
120	40.-	122.-
140	45.-	127.-
160	45.-	134.-
200	53.-	143.-
220	58.-	148.-
240	60.-	153.-
260	63.-	160.-
280	66.-	163.-
300	72.-	168.-

A

A1013 Graphite heat exchangers

Material shell or frame: Steel H II.
Material blocks, pipes or plates: Graphite.
Blocks: Design pressure: 6 bar; design temperature: 160 °C.
Pipe bundle: Design pressure: 6 bar; design temperature: 180 °C.
Plate: Design pressure: 6 bar; design temperature: 180 °C.
Ring nut: Design pressure: 6 bar; design temperature: 180 °C.

Heated surface in m^2	Cost each			
	Blocks	Pipe bundle	Plate	Ring nut
1	13,700.-		12,800.-	10,500.-
2	17,400.-		15,900.-	12,600.-
3	18,800.-	24,400.-	25,300.-	17,400.-
5	31,500.-	28,700.-	28,000.-	24,400.-
10	34,200.-	40,300.-	34,700.-	33,500.-
20	44,500.-	60,900.-	49,200.-	50,000.-
30	67,900.-	79,700.-	78,100.-	
50	93,800.-	115,600.-	99,900.-	
100	138,100.-	180,100.-	162,100.-	

A1014 Basket strainer

Material basket: AISI 316.
Working temperature: 100 °C.
Cast carbon steel: Working pressure 16 bar.
Cast AISI 316: Working pressure 16 bar.
Plate carbon steel: Working pressure 25 bar.

Diameter connection in inches	Cost each			
	Material housing			
	Cast carbon steel	Cast AISI 316	Plate carbon steel	Plate AISI 316
1	320.-	1,200.-	2,500.-	3,540.-
2	490.-	1,910.-	2,580.-	3,720.-
3	720.-	2,700.-	2,880.-	3,920.-
4	1,030.-	3,950.-	3,100.-	4,190.-
6	2,700.-	6,950.-	3,170.-	4,670.-
8	4,450.-		3,610.-	5,180.-
10			4,040.-	6,300.-
12			5,120.-	7,190.-
14			5,720.-	8,840.-

A1015 Filter presses

Material plates: Plastic (polypropylene).
Working pressure: 6 bar.
As from 20 m^2 with automatic plate transport.

Nett filtering surface in m^2	Cost each
2	15,600.-
5	17,800.-
7	19,700.-
10	21,400.-
20	39,000.-
30	47,200.-
50	65,100.-
100	111,400.-

A1016 Plate filters

Working pressure: 6 bar.
Working temperature: Up to 120 °C.
Material plates: AISI 316.

Nett filtering surface in m^2	Cost Each	
	Material filter housing	
	Carbon steel	AISI 316
2	15,900.-	23,800.-
5	17,200.-	25,700.-
7	19,900.-	32,800.-
10	23,000.-	39,200.-
20	28,100.-	49,700.-
30	34,200.-	60,400.-
50	45,400.-	76,800.-
100	84,300.-	135,300.-

A1 Static process equipment

A1017 Candle filters

Material housing	Material candles	Type candles	Nett filtering surface in m²	Cost each
Carbon steel	AISI 316	Wire netting, wrinkled	0.1	400.-
			0.3	1,400.-
			1	2,800.-
			1.5	4,000.-
			2	6,000.-
			2.5	7,000.-
			5	13,400.-
			10	22,600.-
			15	31,800.-
			20	39,100.-
			30	63,100.-
			40	80,600.-
			60	119,600.-
	Disposable candle	Wrinkled	0.1	100.-
			0.3	400.-
			1	1,100.-
			1.5	2,900.-
			2	3,000.-
			2.5	3,100.-
			5	3,500.-
			10	4,300.-
			15	4,600.-
			20	4,900.-
			30	6,200.-
			40	7,300.-
			60	9,500.-
AISI 321	AISI 304	Sintered, cylindrical	0.1	900.-
			0.3	3,100.-
			1	6,800.-
			1.5	10,100.-
			2	13,200.-
			2.5	15,400.-
			5	28,100.-
			10	52,800.-
			15	73,200.-
			20	97,900.-
			30	145,700.-
			40	191,000.-
			60	287,500.-
AISI 316	Disposable candle	Wrinkled	0.1	600.-
			0.3	800.-
			1	3,600.-
			1.5	3,800.-
			2	3,800.-
			2.5	3,900.-
			5	4,500.-
			10	5,100.-
			15	5,500.-
			20	5,900.-
			30	7,800.-
			40	8,900.-
			60	11,000.-

A2 Mechanical process equipment

A2001 Vacuum drum filters

Design:
- Belt discharge, discharge from mills: price x 1.05;
- Pre-coat discharge: price x 1.10;
- Scraper discharge, string discharge.

Excluding:
- Vacuum separators;
- Pumps;
- Piping etc.

Nett filtering surface in m^2	Cost each (x 1,000)	
	Material housing	
	Carbon steel	AISI 316
2	85.-	121.-
5	87.-	124.-
7	92.-	129.-
10	98.-	136.-
20	110.-	149.-
30	132.-	189.-
50	172.-	257.-

A2002 Horizontal vacuum filters for intermittent vacuum

Excluding:
- Vacuum separators;
- Pumps;
- Piping, etc.

Nett filtering surface in m^2	Cost each (x 1,000)	
	Material contact parts	
	AISI 316 L	Plastic
2	101.-	132.-
5	113.-	151.-
7	131.-	174.-
10	152.-	200.-
20	191.-	261.-
30	251.-	339.-
50	365.-	483.-

A2003 Horizontal vacuum filters for continuous vacuum

A

Material rotating parts: AISI 316.
Machines complete assembled and mechanical tested.
Price Ex Works.
Excluding:
- Vacuum separators;
- Pumps;
- Piping, etc.;
- Vacuum station.

Nett filtering surface in m^2	Cost each (x 1,000)
2	159.-
5	176.-
7	213.-
10	233.-
20	304.-
30	344.-
50	534.-
75	637.-
100	645.-

A2004 Water purification plant

Design: Reverse osmosis.
Inlet composition according DIN 2000.
See also water treatment plants.

Permeate capacity in l/h	Efficiency permeate supply	Conductivity		Cost each (x 1,000)
		Inlet in S/cm	Permeate in S/cm	
50	0.5	1,000	50	6.-
100	0.6	1,000	50	8.-
250	0.7	1,000	50	16.-
500	0.75	1,000	20	22.-
1,000	0.75	1,000	20	34.-
2,500	0.75	1,000	20	52.-
5,000	0.8	1,000	20	82.-
10,000	0.8	1,000	10	127.-
25,000	0.8	1,000	10	202.-

A2005 Single-stage centrifugal pumps, 1,450 rpm

Type: Horizontal, single-stage, single suction centrifugal pump with volute, build to pump organic and inorganic fluid.
Material: Fully cast iron, cast steel or AISI 316 (DIN 24256).
Rotational speed: 1,450 rpm.
Base plate: DIN 24259. Mechanical sealing: DIN 24960.
Including:
- Foundation plate;
- Coupling;
- Single mechanical seal;
- Assembly.

Excluding:
- Electric motor;
- Barrier fluid.

Pump size		Pump capacity in m^3/h	Pressure in m	Max. required power in kW	Cost each		
According DIN	In mm				Cast iron	Cast steel	AISI 316
32	125	6.3	5	0.55	3,400.-	4,200.-	5,400.-
	160		8	0.75	3,600.-	4,400.-	5,800.-
	200		12.5	1.5	4,000.-	4,700.-	6,400.-
40	125	12.5	5	0.75	3,500.-	4,300.-	5,600.-
	160		8	1.1	3,700.-	4,400.-	6,000.-
	200		12.5	1.5	4,200.-	5,300.-	6,800.-
	250		20	3	5,400.-	6,700.-	9,300.-
50	125	25	5	1.1	4,000.-	5,000.-	6,400.-
	160		8	2.2	4,400.-	5,400.-	7,200.-
	200		12.5	3	4,700.-	5,800.-	7,700.-
	250		20	4	5,500.-	6,900.-	9,600.-
65	125	50	5	1.5	4,400.-	5,700.-	7,700.-
	160		8	2.2	5,100.-	6,100.-	8,500.-
	200		12.5	4	5,800.-	7,200.-	9,900.-
	250		20	5.5	6,900.-	7,900.-	10,900.-
	315		32	11	8,000.-	9,800.-	13,100.-
80	160	80	8	4	6,100.-	7,400.-	10,000.-
	200		12.5	5.5	7,100.-	8,700.-	11,700.-
	250		20	11	8,100.-	9,800.-	14,200.-
	315		32	18.5	9,400.-	11,600.-	16,200.-
100	200	125	12.5	11	7,900.-	9,400.-	13,100.-
	250		20	15	9,000.-	10,700.-	15,500.-
	315		32	30	8,900.-	10,800.-	14,700.-
	400		50	37	10,600.-	12,700.-	18,100.-
125	250	200	20	22	9,300.-	11,100.-	15,600.-
	315		32	37	10,400.-	12,800.-	17,900.-
	400		50	55	10,600.-	14,500.-	18,400.-
150	315	315	32	45	12,800.-	16,100.-	23,200.-
	400		50	75	13,700.-	18,000.-	24,800.-
250	400	900	40	140		27,700.-	

A2006 Single-stage centrifugal pumps, 2,900 rpm

Type: Horizontal, single-stage, single suction centrifugal pump with volute, build to pump organic and inorganic fluid.
Material: Fully cast iron, cast steel or AISI 316 (DIN 24256).
Rotational speed: 2,900 rpm.
Base plate: DIN 24259. Mechanical sealing: DIN 24960.
Including:
- Foundation plate;
- Coupling;
- Single mechanical seal;
- Assembly.

Excluding:
- Electric motor;
- Barrier fluid.

Pump size		Pump capacity in m^3/h	Pressure in m	Max. required power in kW	Cost each		
According DIN	In mm				Cast iron	Cast steel	AISI 316
32	125	12.5	20	3	3,400.-	4,400.-	5,600.-
	160		32	4	3,700.-	4,500.-	6,000.-
	200		50	11	4,100.-	4,900.-	6,600.-
40	125	25	20	4	3,500.-	4,500.-	5,800.-
	160		32	7.5	4,100.-	5,000.-	6,500.-
	200		50	11	4,400.-	5,800.-	7,600.-
	250		80	18.5	5,500.-	6,900.-	9,500.-
50	125	50	20	7.5	4,100.-	5,300.-	6,700.-
	160		32	11	5,000.-	6,100.-	8,300.-
	200		50	18.5	5,400.-	6,900.-	9,000.-
	250		80	30	6,200.-	8,000.-	11,200.-
65	125	100	20	11	5,100.-	6,600.-	9,000.-
	160		32	15	5,100.-	6,400.-	8,900.-
	200		50	30	5,700.-	7,000.-	9,600.-
	250		80	37	6,600.-	7,800.-	10,700.-
80	160	160	32	22	5,800.-	7,300.-	9,800.-
	200		50	37	5,800.-	7,300.-	10,000.-
	250		80	55	6,900.-	8,500.-	12,100.-
100	200	250	50	55	6,700.-	8,100.-	11,200.-
	250		80	90	7,600.-	9,400.-	13,300.-

A2 Mechanical process equipment

A2007 Single-stage screw compressors

Material: Cast steel.
Outlet pressure: 3.5 bar.
Exclusive electric motor: Compressor without electric motor.
Inclusive electric motor: Electric motor IP 44.
Including:
- Gear box;
- Frame for compressor and electric motor;
- Pressure oil lubricating with pump and cooler;
- Noise reduction outlet side;
- Check valve and pressure safety;
- Coupling and protection cap;
- Instrument panel and attachments.

Max. capacity in m^3/h	Capacity electric motor in kW	Voltage in V	Rotational speed in rpm	Cost compressor	
				Exclusive electric motor	Additional costs electric motor
560	55	380	3,000	29,000.-	3,800.-
1,050	110	380	3,000	43,000.-	8,300.-
2,000	170	3,300	3,000	78,000.-	24,800.-
3,300	260	3,300	3,000	93,000.-	30,500.-
5,100	380	3,300	1,500	147,000.-	41,500.-
8,800	630	3,300	1,500	203,000.-	64,400.-

A2008 Centrifugal fans

Material casing and material impeller.
Including:
- Foundation plate;
- Electric motor.

Material	Pressure in mbar	Rotational speed in rpm	Suction capacity in m^3/h	Cost each
PVC	10	1,500	300	980.-
			600	1,110.-
			900	1,280.-
			1,200	2,100.-
			1,500	2,910.-
Carbon steel plate	30	1,500	2,000	5,570.-
			2,500	5,910.-
			3,000	6,390.-
			4,000	7,070.-
			4,500	7,430.-
			4,800	7,890.-
			5,000	8,220.-
	100	3,000	4,800	6,960.-
			5,000	7,310.-
			5,200	7,890.-
			5,500	7,990.-
			6,000	8,600.-
			6,800	9,540.-
			7,700	12,000.-
			8,700	12,790.-

A2009 Root-blowers

A

Material: Cast iron.
Outlet pressure: ca. 500 mbar.
Design at 1,000 rpm, the electric motor is 3.3 kV.
Including:
- Foundation plate and electric motor;
- Silencer.

Excluding:
- Noise reducing housing.

Rotating speed in rpm	Suction capacity in m^3/h	Cost each
3,000 - 1,500	70	7,100.-
	90	7,700.-
	130	8,100.-
	200	8,400.-
	230	9,600.-
	350	10,000.-
	500	12,000.-
	750	13,800.-
	1,100	15,400.-
	1,500	20,100.-
	2,300	24,300.-
	3,000	29,600.-
	4,700	35,400.-
	7,000	43,400.-
	8,500	46,200.-
1,000.-	9,000	87,800.-
	12,000	105,300.-

A2010 Small mixing vessels

Mixing vessel with propeller mixer with top mounted drive and vertical shaft 60 - 120 rpm drive suited for low viscous fluids.
With motor reductor, suitable for spaces with low explosion risk.
Nett fluid volume: The 'nett' volume in this case nearly matches the volume of the cylindrical part.
Working pressure from -1 bar to +6 bar.

Nett fluid volume in m^3	Cost each (x 1,000)	
	Material	
	Carbon steel	AISI 316
0.25	16.-	29.-
0.5	17.-	34.-
0.8	22.-	39.-
1	23.-	43.-
1.2	24.-	49.-
1.5	27.-	53.-
2	29.-	58.-
3	32.-	68.-
4	35.-	74.-
5	39.-	76.-
10	46.-	99.-
20	61.-	134.-
30	76.-	170.-
70		304.-

A2 Mechanical process equipment

A2011 Mixing vessel with jacket

Mixing vessel with anchor or propeller mixer with top mounted drive and vertical shaft. 60 - 120 rpm drive suited for low viscous fluids. With motor reductor, suitable for spaces with low explosion risk. Working pressure of -1 bar up to +6 bar.
Heated or cooled by jacket
Nett fluid volume: The 'nett' volume in this case nearly matches the volume of the cylindrical part.
Glass enamelled:
- Glass enamelled mixing vessels in accordance with DIN 28136;
- Enamel in accordance with ISO 2743;
- Sliding ring seal with barrier liquid in accordance with DIN 28130;
- Mechanical speed regulation.

AISI 316 L:
- Jacket of steel H III;
- Spherical bottoms;
- Conventional stuffing box;
- Fixed rotation speed.

Additional cost AISI 316 L:
- Sliding ring seal (exclusive barrier liquid facility) € 1,250.- till € 3,100.-.

Nett fluid volume in m^3	Cost each (x 1,000)	
	Material	
	Glass enamelled	AISI 316 L
0.25	40.-	35.-
0.5	53.-	39.-
0.8	55.-	47.-
1	63.-	53.-
1.2	67.-	55.-
1.5	71.-	57.-
2	82.-	60.-
3	95.-	73.-
4	97.-	78.-
5	114.-	81.-
10	161.-	104.-
20	243.-	135.-
30	311.-	178.-

A3 Transportation equipment

A3001 Elevators

Design:
- Model with calibrated chains (DIN 764 or DIN 766);
- Inclusive engine and gear box;
- Steel containers and tension system.

Capacity in m^3	Conveying height in m	Cost each (x 1,000)
30	10	24.-
	20	34.-
	30	50.-
75	10	27.-
	20	44.-
	30	74.-
100	10	33.-
	20	51.-
	30	80.-
250	10	74.-
	20	122.-
	30	186.-
500	10	115.-
	20	168.-
	30	249.-

A3 Transportation equipment

A3002 Belt conveyors

Design:
- Simple version in carbon steel with coating;
- Inclusive engine and gear box;
- Tensioning device;
- Three roll idler;
- Idler spacing top rolls 1.25 m and bottom rolls 5 m;
- Horizontal U-shaped rubber belt.

Variation

Horizontal flat belt.

The price of a flat belt conveyor without spill plates, bulk weight of 1 ton/m^3 and belt speed of 1 m/s, can be 8 - 10% lower than that of the U-shaped variation.

Excluding:
- Assembly;
- Foundation and supporting;
- Chutes, covering, scraper device, alignment system, etc.;
- Central control and surveillance system, emergency brake device.

Belt width in mm	Roller diameter in mm	Length centre to centre pulley in m	Cost per m belt width (c.t.c. pulley)
650	89	25	4,100.-
		50	3,270.-
		75	2,470.-
		100	2,370.-
		150	2,230.-
800	89	25	4,940.-
		50	4,010.-
		75	3,100.-
		100	2,960.-
		150	2,860.-
1,000	108	25	6,830.-
		50	5,580.-
		75	4,340.-
		100	4,010.-
		150	3,710.-
1,200	133	25	8,060.-
		50	7,120.-
		75	6,190.-
		100	5,580.-
		150	4,940.-
		500	4,340.-
1,400	133	25	9,930.-
		50	8,650.-
		75	7,420.-
		100	6,830.-
		150	6,190.-
		500	4,940.-
1,600	133	25	11,160.-
		50	9,930.-
		75	8,650.-
		100	7,740.-
		150	6,830.-
		500	5,930.-
1,800	133	25	12,400.-
		50	11,160.-
		75	9,930.-
		100	9,150.-
		150	8,400.-
		500	7,660.-

A3003 Screw conveyors

A

Design:
- Horizontal set-up;
- Steel trough with end plates;
- Fill and outlet opening;
- Steel conveyor screw.

Excluding:
- Supporting.

Trough length in m	Screw diameter in mm	Cost each
2.5	250	5,600.-
	315	6,000.-
	400	6,500.-
	500	7,100.-
	630	8,300.-
5	250	6,700.-
	315	7,200.-
	400	7,900.-
	500	8,900.-
	630	10,400.-
10	250	11,700.-
	315	12,600.-
	400	14,100.-
	500	17,400.-
	630	21,200.-
20	250	19,900.-
	315	21,900.-
	400	24,300.-
	500	27,700.-
	630	34,300.-

A3004 Extra for screw conveyors drive

Design:
- Electric motor 380/220V - 50 Hz, 1,500 rpm;
- Elastic coupling;
- Inclusive gearbox.

Variation

If reversible additional cost; electric motor, inclusive gearbox plus 10% and screw conveyor plus 3%.

Engine power in kW	Model	Cost each
1.1	With reduction gearbox	430.-
2.2	With reduction gearbox	570.-
5.5	With reduction gearbox	1,230.-
11	With reduction gearbox	2,270.-
22	With reduction gearbox	4,000.-
37	With reduction gearbox	7,030.-

A4 Utilities

A4001 Boiler plants

Central
Complete boiler plants.
Including:
- Chimney;
- Oil facilities;
- Boiler house;
- Insulation.

Excluding:
- Water treatment;
- Oil storage;
- Piping.

Type	Model	Pressure in bar	Temperature in °C	Yield in ton/h	Cost per ton/h steam	
					From	Till
Package unit		30	390	30 - 80	114,200.-	74,800.-
Central	Single unit	30	390	100 - 200	87,800.-	68,300.-
	Multiple units	30	390	100 - 200	134,100.-	99,800.-
	Single unit	100	520	100 - 200	154,000.-	97,600.-

A4002 Water treatment plants

Variation
Depending on water quality and level of automation.
By demineralization also dependent on switching cycle.

Type	Capacity in m^3/h	Cost per m^3/h	
		From	Till
Demineralization-installation	100 - 300	22,500.-	14,700.-
Softening installation		2,200.-	5,100.-
Deferrization installation		2,600.-	5,400.-

A4003 Cooling water-recirculation-unit

Excluding:
- Piping.

Model	Capacity in m^3/h	Cost per m^3/h industrial water	
		From	Till
With cooling tower	1,000 - 5,500	620.-	520.-

A4004 Air compressor installation

Including:
- Buffer tank;
- After-cooler.

Pressure in bar	Output in m^3/h	Cost per m^3/h
8	250.-	710.-
	500.-	490.-
	1,000.-	350.-
	2,000.-	260.-
	4,000.-	170.-

A4005 Instrument air-compressor unit

Type: Oil free.
Including:
- Dryer;
- Buffer tank;
- Oil filter;
- After-cooler.

Pressure in bar	Output in m^3/h	Cost per m^3/h
8	300	820.-
	500	550.-
	1,000	390.-
	2,000	270.-
	4,000	200.-

A5 HVAC equipment

A5001 Central heating unit

Heating unit consist of:
- Heating (gas boiler or heat exchanger);
- Circulation pump;
- Separators;
- Control system;
- Pressure reducing valves;
- Heat distribution systems incl. heating bodies.

Including:
- Insulation;
- Installation work;
- Painting.

Capacity in kW	Type	Cost per kW	
		From	Till
100 - 500	Heat generation	195.-	254.-
	Heat distribution	158.-	297.-

A5002 Cooling unit

Compression installation.
Variation
For industrial cooling installations the costs of special materials and brine refrigerating systems can be a factor 1.5 - 2.5 higher depending on capacity.
Including:
- Distribution pumps;
- Electro motors;
- Internal piping;
- Insulation and painting.

Excluding:
- Structural steel;
- Piping outside unit;
- Power supply cable.

Type	Product	Capacity in MW	Cost per kW	
			From	Till
Compression	Cooled water	500 - 1,000	420.-	320.-
		1,000 - 4,000	320.-	230.-

A5003 Low pressure ventilation system

Including:
- Fans;
- Control system;
- Heating battery;
- Air filters;
- Valve sections;
- Insulation.

Capacity in m^3/h	Cost per m^3 air	
	From	Till
10,000 - 100,000	1.70	4.-

A

A5004 Options for low pressure ventilation system

Part	Cost per m³ air	
	From	Till
Water humidification	0.40	0.80
Steam humidification	0.20	0.30
Mechanical cooling	1.40	1.70

A5005 Ventilating air purification plant

For removing dust and stench (smell) from the ventilating air.

Type filter	Cost per m^3/h	
	From	Till
Carbon filter	11.50	17.80

B1 Pipe and fittings

B1001 Seamless steel pipes acc ASME/ANSI, (Std) (40)

Pipe type: ASTM-A106.
Average prices for new build projects with a volume of max. 5,000 tons of piping. Project discount is included. For larger volumes, there is a volume discount. By small volumes (modifications, plant site projects), prices can be significantly higher.

Variation

Factor for seamless steel pipes by DIN:
- Nominal pipe size < 100 mm (< 4"): 0.85;
- Nominal pipe size > 100 mm (> 4"): 1.15.

Nominal diameter in inches	External diameter in mm	Wall thickness in mm	Weight in kg/m	Cost per 100 m
0.5	21.3	2.8	1.27	277.-
0.75	26.7	2.9	1.68	338.-
1	33.4	3.4	2.5	461.-
1.5	48.3	3.7	4.05	707.-
2	60.3	3.9	5.44	738.-
3	88.9	5.5	11.3	1,310.-
4	114.3	6	16.1	1,805.-
5	141.3	6.6	21.8	2,615.-
6	168.3	7.1	28.3	3,360.-
8	219.1	8.2	42.5	5,165.-
10	273	9.3	60.3	7,450.-
12	323.9	9.5	73.8	9,350.-
14	355.6	9.5	81.3	11,235.-
16	406.4	9.5	93.3	13,020.-
20	508	9.5	117	16,490.-

B1 Pipe and fittings

B1002 Carbon steel flanges and fittings by ASME/ANSI, (Std)

Welding neck flange, 150 lbs: ASTM-A105.
Welding neck flange, 300 lbs: ASTM-A105.
Elbow, 90°: ASTM-A234 WPB.
Tee, d × d × d: ASTM-A234 WPB.
Concentric reducer: ASTM-A234 WPD, reducing to the adjacent lower diameter.
Average prices for new build projects with a volume of max. 5,000 tons of piping. Project discount is included. For larger volumes, there is a volume discount. By small volumes (modifications, plant site projects), prices can be significantly higher.

Remarks

DIN Unit prices.
Price recalculation factor for fittings is 1.0 according DIN.

B PIPING MATERIAL

B1 Pipe and fittings

B1002 Carbon steel flanges and fittings by ASME/ANSI, (Std)

Part	Nominal diameter in inches	Cost each	
		150 lbs	300 lbs
Welding neck flange	0.5	3.15	4.75
	0.75	3.70	5.25
	1	4.75	5.80
	1.5	5.80	10.-
	2	6.85	11.05
	3	10.50	14.70
	4	14.20	23.65
	5	20.70	35.20
	6	24.15	45.70
	8	35.70	65.85
	10	56.20	91.90
	12	81.40	140.-
	14	108.-	181.-
	16	141.-	241.-
	20	183.-	388.-
Elbow	0.5	1.60	
	0.75	2.10	
	1	2.10	
	1.5	3.70	
	2	4.75	
	3	8.40	
	4	13.65	
	5	26.80	
	6	34.65	
	8	48.85	
	10	86.10	
	12	130.-	
	14	175.-	
	16	223.-	
	20	359.-	
T-piece	0.5	2.65	
	0.75	2.65	
	1	4.20	
	1.5	7.35	
	2	9.45	
	3	13.65	
	4	23.10	
	5	36.75	
	6	51.45	
	8	65.10	
	10	110.-	
	12	180.-	
	14	261.-	
	16	302.-	
	20	516.-	
Concentric reducer	0.75	1.05	
	1	2.10	
	1.5	3.15	
	2	3.70	
	3	5.25	
	4	7.35	
	5	11.55	
	6	15.75	
	8	19.95	
	10	33.10	
	12	45.70	
	14	84.55	
	16	111.-	
	20	204.-	

B PIPING MATERIAL

B1 Pipe and fittings

B1003 Seamless steel pipes by ASME/ANSI, (XS)

Pipe type: ASTM-A106
Average prices for new build projects with a volume of max. 5,000 tons of piping. Project discount is included. For larger volumes, there is a volume discount. By small volumes (modifications, plant site projects), prices can be significantly higher.

Remarks

DIN price per unit for pipe fittings.
For conversion of pipe fitting prices according DIN the factor is 1.0.

B

B1 Pipe and fittings

B1003 Seamless steel pipes by ASME/ANSI, (XS)

Nominal diameter in inches	Outside diameter in mm	Wall thickness in mm	Weight in kg/m	Cost per 100 m
0.5	21.3	3.7	1.62	315.-
0.75	26.7	3.9	2.2	389.-
1	33.4	4.5	3.24	536.-
1.5	48.3	5.1	5.41	840.-
2	60.3	5.5	7.48	1,060.-
3	88.9	7.6	15.3	1,755.-
4	114.3	8.6	22.3	2,635.-
5	141.3	9.5	31	3,770.-
6	168.3	11	42.6	4,895.-
8	219.1	12.7	64.6	8,010.-
10	273	12.7	81.6	10,285.-
12	323.9	12.7	97.5	12,375.-
14	355.6	12.7	107	14,905.-
16	406.4	12.7	123.16	17,045.-
20	508	12.7	155	21,235.-

B1004 Carbon steel flanges and fittings by ASME/ANSI, (XS)

Welding neck flange, 150 lbs: ASTM-A105.
Welding neck flange, 300 lbs: ASTM-A105.
Elbow, 90°: ASTM-A234 WPB.
Tee, d × d × d: ASTM-A234 WPB.
Concentric reducer: ASTM-A234 WPD, reducing to the adjacent lower diameter.
Average prices for new build projects with a volume of max. 5,000 tons of piping. Project discount is included. For larger volumes, there is a volume discount. By small volumes (modifications, plant site projects), prices can be significantly higher.

Remarks
DIN Unit prices.
For conversion of pipe fitting prices according DIN, factor is 1.0.

B1004 Carbon steel flanges and fittings by ASME/ANSI, (XS)

Part	Nominal diameter in inches	Cost each	
		150 lbs	300 lbs
Welding neck flange	0.5	4.-	4.25
	0.75	5.10	5.55
	1	6.-	8.35
	1.5	7.35	10.75
	2	9.50	13.-
	3	11.95	16.75
	4	18.25	26.-
	5	24.45	41.70
	6	30.65	57.40
	8	49.40	81.30
	10	69.35	128.-
	12	108.-	175.-
	14	147.-	198.-
	16	171.-	276.-
	20	220.-	459.-
Elbow	0.5	2.45	
	0.75	2.95	
	1	3.40	
	1.5	5.15	
	2	7.40	
	3	11.-	
	4	23.-	
	5	35.20	
	6	47.20	
	8	80.-	
	10	123.-	
	12	209.-	
	14	263.-	
	16	441.-	
	20	769.-	
Tee	0.5	4.-	
	0.75	6.-	
	1	6.20	
	1.5	10.80	
	2	15.-	
	3	20.-	
	4	27.50	
	5	50.50	
	6	68.30	
	8	116.-	
	10	164.-	
	12	233.-	
	14	376.-	
	16	647.-	
	20	935.-	
Concentric reducer	0.75	2.75	
	1	3.65	
	1.5	5.45	
	2	7.25	
	3	6.30	
	4	9.-	
	5	17.25	
	6	25.25	
	8	43.50	
	10	61.80	
	12	60.85	
	14	105.-	
	16	186.-	
	20	387.-	

B1 Pipe and fittings

B1005 Welded stainless steel pipes

Pipe according to ANSI B36.19.
304L: welded stainless steel pipe (ANSI-Schedule 10S).
316L: welded stainless steel pipe (ANSI-Schedule 10S).
Average prices for new build projects with a volume of max. 5,000 tons of piping. Project discount is included. For larger volumes, there is a volume discount. By small volumes (modifications, plant site projects), prices can be significantly higher.

Variation

Factor for welded stainless steel pipes by DIN:
- Nominal pipe size < 100 mm (= 4"): 0.85;
- Nominal pipe size > 100 mm (> 4"): 1.15.

Nominal diameter in inches	Outside diameter in mm	Wall thickness in mm	Weight in kg/m	Cost per m	
				304 L	316 L
0.5	21.34	2.11	0.99	10.65	13.-
0.75	26.67	2.11	1.27	14.-	18.-
1	33.4	2.77	2.08	18.-	23.-
1.25	42.16	2.77	2.69	22.15	26.-
1.5	48.26	2.77	3.1	24.-	28.-
2	60.32	2.77	3.92	24.-	34.-
2.5	73.02	3.05	5.25	32.-	40.-
3	88.9	3.05	6.44	40.-	49.-
4	114.3	3.05	8.34	57.-	71.-
5	141.3	3.4	11.56	75.-	93.-
6	168.27	3.4	13.82	93.-	115.-
8	219.07	3.76	19.92	104.-	133.-
10	273.05	4.19	27.75	142.-	166.-
12	323.85	4.57	35.96	174.-	204.-

B PIPING MATERIAL

B1 Pipe and fittings

B1006 Stainless steel fittings

304 L: fittings for stainless steel pipes (ANSI-Schedule 10S).
316 L: fittings for stainless steel pipes (ANSI-Schedule 10S).
Pipe according to ANSI B16.9.
Stub-end: short.
Elbow: 90°, r 1.5 d.
Tee: dxdxd.
Average prices for new build projects with a volume of max. 5,000 tons of piping. Project discount is included. For larger volumes, there is a volume discount. By small volumes (modifications, plant site projects), prices can be significantly higher.
Remarks
DIN Unit prices.
For conversion of pipe fitting prices according DIN the factor is 1.0.

Part	Nominal diameter in inches	Cost each	
		304 L	316 L
Stub-end	0.5	4.-	5.-
	0.75	5.-	7.-
	1	10.-	14.-
	1.25	12.-	17.50
	1.5	14.-	21.-
	2	27.-	35.-
	2.5	39.-	47.-
	3	45.-	59.-
	4	51.-	76.-
	5	72.-	95.50
	6	93.-	115.-
	8	116.-	159.-
	10	169.-	267.-
	12	213.-	357.-
Elbow	0.5	3.-	4.-
	0.75	5.50	10.-
	1	8.-	15.25
	1.25	13.50	21.-
	1.5	19.-	27.-
	2	23.65	32.-
	2.5	38.-	47.-
	3	42.-	62.-
	4	60.50	78.-
	5	91.25	110.-
	6	122.-	142.-
	8	175.-	196.-
	10	300.-	346.-
	12	435.-	571.-
Tee	0.5	9.-	10.-
	0.75	11.-	12.-
	1	10.-	18.-
	1.25	19.-	22.-
	1.5	26.-	38.35
	2	25.55	34.65
	2.5	44.-	44.95
	3	61.95	55.20
	4	79.65	96.-
	5	109.-	130.-
	6	134.-	164.-
	8	192.-	253.-
	10	298.-	406.-
	12	543.-	667.-

B1007 Hard PE pipe (HPE-80)

Dimensions by DIN 8074/75.
Working pressure at 20 °C.

Outside diameter in mm	Wall thickness in mm	Working pressure in bar	Cost per 100 m
20	1.9	10	215.-
25	1.6	6	205.-
	2.3	10	268.-
32	1.9	6	205.-
	3	10	268.-
40	2.3	6	278.-
	3.7	10	436.-
50	2.9	6	467.-
	4.6	10	683.-
63	3.6	6	698.-
	5.8	10	1,050.-
75	4.3	6	977.-
	6.9	10	1,510.-
90	5.1	6	1,375.-
	8.2	10	2,035.-
110	6.3	6	2,035.-
	10	10	2,980.-
125	7.1	6	2,490.-
	11.4	10	3,845.-
160	9.1	6	4,190.-
	14.6	10	6,195.-
200	11.4	6	6,405.-
	18.2	10	9,765.-
250	14.2	6	9,765.-
	22.8	10	15,225.-
315	17.9	6	16,380.-
	28.7	10	25,095.-

B PIPING MATERIAL

B1 Pipe and fittings

B1008 Hard PVC pipe

Dimensions by DIN 8061/62.
Working pressure at 20 °C.

Outside diameter in mm	Wall thickness in mm	Working pressure in bar	Cost per 100 m
25	1.5	10	189.-
32	1.8	10	221.-
40	1.9	10	284.-
50	1.8 2.4	6 10	336.- 404.-
63	1.8 3	6 10	483.- 599.-
75	2.2 3.6	6 10	662.- 872.-
90	2.7 4.3	6 10	1,070.- 1,240.-
110	3.2 5.3	6 10	1,365.- 1,805.-
125	3.7 6	6 10	1,975.- 2,520.-
160	4.7 7.7	6 10	3,160.- 3,820.-
200	5.9 9.6	6 10	5,565.- 6,775.-
250	7.3	6	8,410.-

B1009 Glass fibre reinforced epoxy pipe (GRE)

Working pressure at 110 °C.
Glued connections.

Nominal diameter in mm	Working pressure in bar	Length pipe in m	Cost per m
25	16	3	20.-
40	16	6	26.-
50	16	6	29.-
80	16	6	37.-
100	16	6	46.-
150	16	10	55.-
200	16	10	70.-
250	16	10	88.-
300	16	10	106.-

B

B PIPING MATERIAL

B1 Pipe and fittings

B1010 Fittings and flanges for glass fibre reinforced epoxy pipe (GRE)

Working pressure at 110 °C.
Glued connections.
Concentric reducer: reducing to the adjacent lower diameter.

Pipe fitting	Nominal diameter in mm	Working pressure in bar	Cost each
Pipe bend, 45°	25	32	31.-
	40	32	34.-
	50	32	38.-
	80	32	43.-
	100	25	55.-
	150	20	69.-
	200	16	101.-
	250	12.5	135.-
	300	12.5	188.-
Pipe bend, 90°	25	32	51.-
	40	32	55.-
	50	32	67.-
	80	32	79.-
	100	25	85.-
	150	20	112.-
	200	16	173.-
	250	12.5	197.-
	300	12.5	260.-
Tee (d x d x d)	25	32	108.-
	40	32	124.-
	50	32	135.-
	80	32	147.-
	100	25	203.-
	150	20	255.-
	200	16	329.-
	250	12.5	430.-
	300	12.5	550.-
Reducer	40	32	37.-
	50	32	38.-
	80	32	40.-
	100	25	42.-
	150	20	84.-
	200	16	103.-
	250	12.5	137.-
	300	12.5	202.-
Flange	25	32	52.-
	40	32	61.-
	50	32	64.-
	80	32	87.-
	100	25	95.-
	150	20	116.-
	200	16	142.-
	250	12.5	164.-
	300	12.5	203.-

B2 Ducting en chimneys

B2001 Spiral fabricated pipe

Type: Galvanized.
Length: Standard length is 6 m.
Standard diameter and wall thickness.

Nominal diameter in mm	Wall thickness in mm	Cost per m
80	0.5	5.10
100	0.5	7.20
125	0.5	8.20
150	0.5	9.30
160	0.5	9.30
200	0.5	12.30
250	0.5	14.40
300	0.6	19.60
315	0.6	21.60
400	0.6	26.80
500	0.8	41.20
630	0.8	53.50
800	1.0	81.30

B2 Ducting en chimneys

B2002 Fittings for spiral piping

Type: Galvanized.
Bent segments: 90°, r 1.5 d.
Tee: d × d × d.
Concentric reducer: Reducing to the adjacent lower diameter.

Part	Nominal diameter in mm	Cost each
Bent segments	80	9.45
	100	10.50
	125	14.70
	150	16.80
	160	21.-
	200	46.20
	250	50.40
	300	64.05
	315	65.10
	400	78.75
	500	102.-
	630	186.-
	800	415.-
Tee	80	10.50
	100	11.55
	125	12.60
	150	18.90
	160	19.95
	200	26.25
	250	36.75
	300	42.-
	315	45.15
	400	65.10
	500	105.-
	630	218.-
	800	333.-
Reducer	100	10.50
	125	11.55
	150	12.60
	160	14.70
	200	15.75
	250	19.95
	300	22.05
	315	24.15
	400	37.80
	500	47.25
	630	70.35
	800	126.-
Plug and socket connection	80	3.15
	100	3.15
	125	3.15
	150	3.15
	160	3.15
	200	3.15
	250	4.20
	300	5.25
	315	5.25
	400	6.30
	500	10.50
	630	16.80
	800	30.45

B2 Ducting en chimneys

B2003 Square shaped air ducts

Ducting, fittings, fasteners.
Duct surface: circumference x length of duct.
Including:
- Assembly.

Material	Costs per m^2 duct surface
Galvanized steel plate	80.60
AISI 304	166.-
Aluminium	102.-

B2004 Insulation for square shaped air ducting

Insulation of ducts.
Duct surface: circumference x length of duct.
Including:
- Assembly.

Location	Material	Thickness in mm	Cost per m^2 duct surface
Inside	Linecoustic	40	60.-
Outside	Thermoflex	15	16.80
	Mineral wool with aluminium foil	40	21.-
	Mineral wool with aluminium plate	40	95.-

B2005 Chimneys

Material and prefabrication.
Price based on the weight of the chimney.
Bottom part generally 5 mm thick.
Supporting: blasted and one coating.

Part	Material	Cost per kg
Chimney	Corten steel	11.60
	AISI 304	18.40
Steel construction	Carbon steel	8.20

B3 Valves and appendages

B3001 Ball valves, ANSI, cast iron

Type: with flanges.
ANSI, reduced bore, pressures acc ANSI-B 16.5.
Materials
Housing: cast iron.
Ball: AISI 316.
Sealing: PTFE.
Spindle: AISI 316.
Variation
Conversion to DIN prices: factor 1.25.

Nominal diameter in inches	Cost each	
	150 lbs	300 lbs
0.5	102.-	128.-
0.75	112.-	143.-
1	128.-	163.-
1.5	204.-	281.-
2	240.-	347.-
3	423.-	648.-
4	638.-	898.-
6	1,665.-	2,490.-
8	3,280.-	3,930.-

B3002 Ball valves, ANSI, AISI 316

Type: with flanges.
ANSI, reduced bore, pressures acc ANSI-B 16.5.
Materials
Housing: AISI 316.
Ball: AISI 316.
Sealing: PTFE.
Spindle: AISI 316.
Variation
Conversion to DIN prices: factor 1.25.

Nominal diameter in inches	Cost each	
	150 lbs	300 lbs
0.5	140.-	170.-
0.75	160.-	200.-
1	190.-	240.-
1.5	310.-	390.-
2	410.-	520.-
3	790.-	1,050.-
4	1,500.-	1,820.-
6	2,780.-	3,880.-
8	4,710.-	7,330.-

B3003 Check valves, ANSI

Type: ANSI with flanges.
Pressure stages by ANSI-B 16.5.
Materials
Housing: cast iron.
Housing: cast iron.
Variation
Conversion to DIN prices: factor 1.25.

Nominal diameter in inches	Cost each	
	150 lbs	300 lbs
2	184.-	301.-
3	265.-	464.-
4	367.-	653.-
6	556.-	933.-
8	847.-	1,220.-
10	1,335.-	2,040.-
12	2,355.-	3,745.-

B3004 Butterfly valves, ANSI

Type: ANSI, double flanged.
Pressures acc ANSI-B 16.5.
Materials
Housing: cast iron.
Valve: cast iron.
Valve sealing: Buna-N.
Pressure: 150 lbs.
Variation
Conversion to DIN prices: factor 1.25.

Nominal diameter in		Cost each
Inches	mm	
2	50	154.-
4	100	230.-
6	150	369.-
8	200	520.-
10	250	834.-
12	300	943.-
14	350	1,265.-
16	400	1,770.-
20	500	3,470.-
24	600	6,030.-
28	700	10,710.-
30	750	13,160.-
32	800	15,650.-
36	900	18,820.-
40	1,000	24,650.-
42	1,050	33,750.-
48	1,200	40,820.-

B3005 Plug valves, ANSI, straight model

Materials
Plug: CrNi.
Housing: Cast iron.
Bush: PTFE.
Type: Flanged.
Wrench operated up to 4 inches.
Variation
Conversion to DIN prices: factor 1.25.

Nominal diameter in inches	Cost each	
	150 lbs	300 lbs
1	199.-	237.-
1.5	254.-	329.-
2	376.-	520.-
3	474.-	636.-
4	554.-	922.-

B3006 Membrane valves with female thread connection, BSP

Housing with female thread connection (BSP), fitted with an EP rubber membrane, a hand operated cap with rising hand wheel and position indicator.
Variation
At lower/additional cost are other membrane types available (PTFE). Other hand operated and pneumatic actuators are also available.

Nominal diameter in mm	Cost each		
	Housing material		
	Cast iron GJS-450-10	Bronze LG2	Stainless steel 316C16
15	45.-	60.-	138.-
20	53.-	76.-	210.-
25	58.-	81.-	215.-
32	69.-	139.-	388.-
40	90.-	149.-	398.-
50	108.-	198.-	495.-
65		295.-	837.-
80		333.-	873.-

B3007 Membrane valves with flange connection

Housing with flange connection (EN1092-1/2) type B,PN10/16.
Unlined and rubber lined valves with EP rubber membrane.
Glass-, ETFE-, or PFA lined valves with a PTFE/EPM membrane.
Hand operated cap with rising hand wheel and position indicator.

Variation

At lower/additional cost are other membrane types available (PTFE). Other hand operated and pneumatic operated caps are also available. Also, other flange connections available (for instance ANSI 150 lbs)

Nominal diameter in mm	Cost each						
	Housing material						
	Cast iron GJL-250, unlined	Nodular iron GJS-400-18-LT, unlined	Stainless steel 316C16 unlined	Cast iron GJL-250 ebonite lined (hard rubber)	Cast iron GJL-250 glass lined	Nodular iron GJS-400-18-LT, EFTE lined	Cast iron GJS-400-18-LT
15	85.-	148.-	295.-	Not available	Not available	Not available	Not available
20	96.-	152.-	326.-	190.-	341.-	352.-	391.-
25	99.-	156.-	338.-	190.-	349.-	357.-	397.-
32	116.-	216.-	552.-	222.-	541.-	536.-	514.-
40	138.-	227.-	557.-	241.-	573.-	541.-	600.-
50	159.-	238.-	638.-	305.-	703.-	665.-	740.-
65	203.-	406.-	1,075.-	412.-	1,040.-	1,015.-	1,005.-
80	264.-	449.-	1,115.-	455.-	1,125.-	1,125.-	1,220.-
100	396.-	655.-	1,895.-	685.-	1,915.-	1,890.-	2,010.-
150	723.-	1,795.-	3,900.-	1,360.-	3,935.-	3,990.-	Not available

B3008 Gate valves, type full bore

Type: Full bore.
150 and 300 lbs: Flange connection.
600 and 900 lbs: Weld connection.

Variation

Conversion to DIN prices: factor 1.25.

Material	Nominal diameter in inches	Cost each			
		150 lbs	300 lbs	600 lbs	900 lbs
Cast steel	2	229.-	232.-	463.-	863.-
	3	327.-	361.-	529.-	1,180.-
	4	492.-	597.-	730.-	2,175.-
	6	561.-	915.-	1,670.-	4,490.-
	8	1,320.-	1,290.-	2,870.-	8,810.-
	10	1,105.-	1,585.-	4,320.-	11,140.-
	12	1,645.-	2,290.-	5,640.-	15,770.-
AISI 316	2	580.-	754.-	1,170.-	
	3	783.-	996.-	1,780.-	
	4	1,150.-	1,450.-	2,550.-	
	6	1,890.-	2,890.-	4,720.-	
	8	3,290.-	5,100.-	10,030.-	
	10	6,050.-	7,690.-	15,290.-	

B3 Valves and appendages

B3009 Gate valves, type reduced bore

Thread or weld connection.
Variation
Conversion to DIN prices: factor 1.25.

Material	Nominal diameter in inches	Cost each	
		800 lbs	1,500 lbs
Cast iron	0.25	98.80	
	0.375	102.-	
	0.5	104.-	172.-
	0.75	106.-	176.-
	1	124.-	202.-
	1.5	200.-	366.-
	2	240.-	465.-
AISI 316	0.5	127.-	282.-
	0.75	189.-	353.-
	1	248.-	742.-
	1.5	593.-	1,260.-
	2	643.-	1,585.-

B3010 Safety valves and pressure relief valves

Type: Flange connection.

Housing material	Connection in inches		Cost each		
	Inlet	Outlet	150 lbs	300 lbs	600 lbs
Cast steel	1	1.5	949.-	1,000.-	1,325.-
	1.5	2	1,255.-	1,305.-	1,735.-
	2	3	1,655.-	1,815.-	2,410.-
	3	4	1,695.-	1,940.-	2,580.-
	4	6	2,530.-	2,835.-	3,765.-
	6	8	4,520.-	5,215.-	6,915.-
Stainless steel	1	1.5	2,020.-	2,070.-	2,755.-
	1.5	2	2,090.-	2,205.-	2,920.-
	2	3	2,665.-	2,775.-	3,685.-
	3	4	3,030.-	3,530.-	4,670.-
	4	6	5,325.-	6,285.-	8,335.-
	6	8	6,700.-	6,985.-	10,345.-

B3 Valves and appendages

B3011 Rupture disc holders and rupture discs

Rupture disc
Material of rupture disc depends on pressure and temperature. Stainless steel or nickel material. Lined or unlined. Tolerance +5/-10%.
Design at least four rupture discs per holder in relation to production costs per application.

Part	Material	Connection in mm	Cost each
Holder	Cast steel	Up to 40	321.-
		50	505.-
		80	546.-
		100	750.-
		150	1,110.-
		200	1,340.-
	Stainless steel	40	479.-
		50	755.-
		80	806.-
		100	1,110.-
		150	1,640.-
		200	2,025.-
Rupture disc	Stainless steel or nickel	40	245.-
		50	270.-
		80	321.-
		100	398.-
		150	556.-
		200	653.-

B4 Complete delivered and installed piping systems

B4001 Complete assembled underground piping system

Prices are based on a length of at least 10 km in a relatively open outside infrastructure.
Variation
Prices are strongly influenced by:
- Depth;
- Geography;
- Soil type;
- Number of road and water crossings;
- Insulation;
- Working pressure;
- Agriculture and land situation;
- Material prices.

Including:
- Indirect costs (construction management), general overhead cost;
- Design (pressure 60 bar, temperature 50 °C, no thermal insulation);
- Compensation of landowners and landusers;
- Materials and erection;
- Testing and commissioning;
- Temporary building facilities;
- Small road- and water crossings (2 drills of 400 meter each);
- Ground cover 1.25 meter.

Excluding:
- Big river-, channel- and road crossings;
- PIG launcher, receiving station and block valves;
- Licences and tax;
- Owners internal costs.

Diameter in inches	Cost per m CS		Cost per m duplex	
	From	Till	From	Till
4	265.-	468.-	151.-	671.-
6	312.-	556.-	520.-	967.-
8	354.-	645.-	671.-	1,260.-
10	411.-	733.-	816.-	1,545.-
12	468.-	848.-	1,005.-	1,870.-
16	489.-	900.-		
20	551.-	1,015.-		
24	645.-	1,210.-		
30	884.-	1,695.-		
36	1,145.-	2,200.-		

B4 Complete delivered and installed piping systems

B4002 Specifications for complete fabricated and installed piping

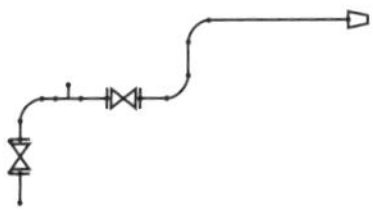

Type A

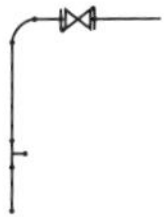

Type B

Material
Both types based on 10 meter pipe length.
Including:
- Fabrication and installation.

Excluding:
- Cost of valves;
- Insulation and painting;
- Supporting;
- Pipe bridge;
- Scaffolding;
- Design costs.

Part	Unit	Quantity	
		Type A	Type B
Pipe	m	11.-	11.-
Welds	piece	16.-	9.-
Elbow	piece	4.-	2.-
Tee	piece	2.-	1.-
Concentric reducer	piece	2.-	1.-
Welding neck flange	piece	4.-	2.-
Valve (only installation cost)	piece	2.-	1.-

B4003 Complete assembled piping system ANSI 150 lbs

Type: Welded pipe.
Remarks
For pipe lengths 200 - 500 meter, a cost reduction of 5 - 15%.
For type B a reduction of 40% for large straight lengths in sleepers or pipe bridges. For more specifications refer to B4002.
Conversion to DIN prices: factor 0.95

Nominal diameter		Cost per m	
Inches	mm	Type A	Type B
0.5	15	175.-	121.-
0.75	20	196.-	131.-
1	25	217.-	148.-
1.25	32	239.-	163.-
1.5	40	266.-	181.-
2	50	273.-	197.-
2.5	65	328.-	237.-
3	80	384.-	278.-
4	100	479.-	317.-
5	125	593.-	391.-
6	150	707.-	465.-
8	200	977.-	634.-
10	250	1,390.-	899.-
12	300	1,820.-	1,165.-

B4 Complete delivered and installed piping systems

B4004 Complete assembled piping system ANSI 300 lbs

Type: Welded pipe.
Remarks
For pipe lengths 200 - 500 meter, a cost reduction of 5 - 15%.
For type B a reduction of 40% for large straight lengths in sleepers or pipe bridges. For more specifications refer to B4002.
Conversion to DIN prices: factor 0.95.

Nominal diameter		Cost per m	
Inches	mm	Type A	Type B
0.50	15	182.-	120.-
0.75	20	203.-	131.-
1	25	224.-	145.-
1.25	32	250.-	160.-
1.50	40	275.-	175.-
2	50	303.-	191.-
2.50	65	364.-	228.-
3	80	426.-	265.-
4	100	530.-	328.-
5	125	655.-	404.-
6	150	780.-	480.-
8	200	1,075.-	653.-
10	250	1,525.-	925.-
12	300	2,000.-	1,200.-

B4005 Complete fabricated and installed jacketed pipe

Type
Inside pipe: AISI 304, seamless by DIN 2642.
Outside pipe: Carbon steel, DIN 17175.
More specifications concerning type A and B refer to B4002.
Variation
Outside diameter is 2" larger than inside diameter.

Diameter in inches	Cost per m	
	Type A	Type B
2	1,310.-	793.-
2.5	1,615.-	927.-
3	1,605.-	979.-
4	1,750.-	1,080.-
5	2,100.-	1,380.-
6	2,175.-	1,400.-
8	2,730.-	1,740.-

B4006 Steam tracing, spiral wound or parallel tracing, sched 80

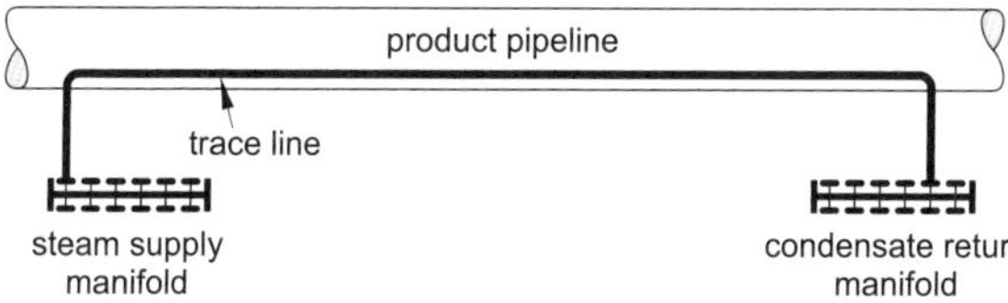

Diameter of pipe: DN 50 - DN 250.
Specifications of type A and B see B.4.002.

Variation

Installation of line from manifold valve to condenser valve including bending, welding, testing and strapping. Excluding prefabrication and assembly of manifolds, steam supply and condensate return lines.

Including:
- Couplings:
- Assembly of trace lines.

Excluding:
- Valves;
- Steam traps;
- Steam supply lines and condensate lines;
- Insulation.

Type of tracing	Diameter	Type	Cost per m heat tracing	
			From	Till
Parallel tracing (1 strand)	0.5" - 0.75"	A	118	171
		B	72	128
Winding tracing	0.5" - 0.75"	A	91	144
		B	70	102

B4007 Complete fabricated and installed stainless steel piping, DIN PN 10, AISI 304L

Type: Welded pipe.
Remarks
For pipe lengths 200 - 500 meter, a cost reduction of 5 - 15%.
For type B a reduction of 40% for large straight lengths in sleepers or pipe bridges. For more specifications refer to B4002.

Nominal diameter		Cost per m	
Inches	mm	Type A	Type B
0.50	15	211.-	137.-
0.75	20	241.-	154.-
1	25	267.-	196.-
1.25	32	299.-	223.-
1.50	40	331.-	214.-
2	50	379.-	249.-
2.50	65	476.-	320.-
3	80	572.-	391.-
4	100	723.-	496.-
5	125	893.-	622.-
6	150	1,065.-	747.-
8	200	1,365.-	878.-
10	250	1,755.-	1,140.-
12	300	2,205.-	1,420.-

B4008 Complete fabricated and installed stainless steel piping, DIN PN 10, AISI 316L

Type: Welded pipe.
Remarks
For pipe lengths 200 - 500 meter, a cost reduction of 5 - 15%.
For type B a reduction of 40% for large straight lengths in sleepers or pipe bridges. For more specifications refer to B4002.

Nominal diameter		Cost per m	
Inches	mm	Type A	Type B
0.50	15	217.-	142.-
0.75	20	250.-	161.-
1	25	281.-	181.-
1.25	32	319.-	206.-
1.50	40	357.-	231.-
2	50	409.-	264.-
2.50	65	504.-	328.-
3	80	598.-	391.-
4	100	771.-	503.-
5	125	995.-	646.-
6	150	1,220.-	788.-
8	200	1,630.-	1,075.-
10	250	2,160.-	1,430.-
12	300	2,785.-	1,825.-

B4009 Specifications for complete installed pipe systems with Victaulic couplings, parts

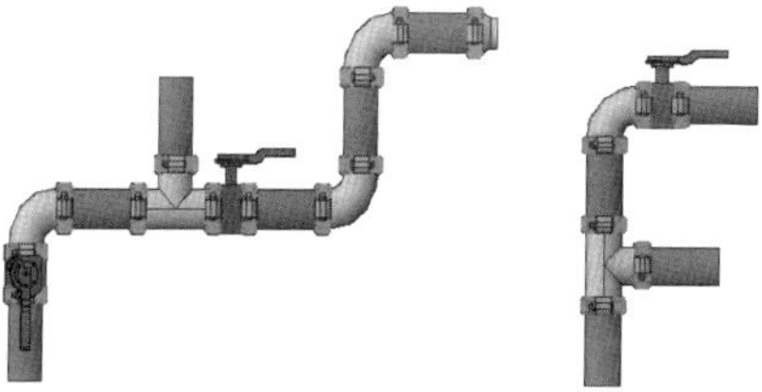

Material
Based on 10 meter pipe length.
Including:
- Supply of material, fabrication and installation.

Excluding:
- Cost of valves;
- Insulation and painting;
- Supporting;
- Pipe bridge;
- Scaffolding;
- Design costs.

Part	Unit	Quantity	
		Type A	Type B
Pipe	m	10	10
Couplings	piece	12	6
Elbow	piece	3	1
T-piece	piece	1	1
Concentric reducer	piece	1	0
Valve (only installation costs)	piece	2	1

B4010 Specifications for complete installed pipe systems with Victaulic couplings

Material
Based on 10 meter pipe length.
Including:
- Supply of material, fabrication and installation.

Excluding:
- Cost of valves;
- Insulation and painting;
- Supporting;
- Pipe bridge;
- Scaffolding;
- Design costs.

Nominal diameter in inches	Cost per 10 m	
	Type A	Type B
3	3,830.-	2,130.-
4	4,820.-	2,700.-
6	7,120.-	3,920.-
8	9,140.-	5,080.-
10	13,620.-	7,620.-
12	19,100.-	10,640.-

B4 Complete delivered and installed piping systems

B4011 Specifications for complete installed piping with victaulic couplings

Nominal diameter in inches	Cost per m	
	Type A	Type B
2	105.-	75.-
2.5	125.-	90.-
3	145.-	110.-
4	200.-	145.-
5	285.-	205.-
6	345.-	250.-
8	510.-	360.-
10	725.-	505.-
12	870.-	605.-
14	1,290.-	795.-
16	1,600.-	970.-
18	1,920.-	1,150.-
20	2,350.-	1,405.-
24	3,445.-	1,890.-

B

C PAINTING AND INSULATION

C1 Painting

C1001 Surface treatment

Excluding:
Rust removal in the field of existing small surfaces mechanically by hand tools (hourly rate)

Activity	Surface treatment	Cost per m^2
Treatment of steel	Blasting sa-2.5 -shop	13.-
	Blasting sa-2.5 -field	62.-
	Blasting sa-3 -field	73.-

C1002 Painting on steel

Painting products which are fluid and based on solvent- or water, free from solvent.
Rates are based on shop work.
Excluded are transport damages or field touch-up (hourly rate)

Painting	Type of layer	Cost per m^2
Applying wet paint	Primer	16.-
	Intermediate coat	14.-
	Top coat	12.-

C

C2 Insulation

C2001 Thermal insulation of equipment

Cost per square meter. The total costs are defined by the sum of all calculated surface areas. The surface area is calculated over the exterior of the insulation. The body runs from tangent line to tangent line.
Other parts like heads, bottom, caps, manholes, support and vacuum rings are calculated using a factor. Overlaps and spaces less than 1 m 2 are not included in the calculations.
The surface area of connections (nozzles) and supports must be calculated separately. The height is defined as the distance from the exterior of the equipment to the outside layer of the insulation (see also comments on measurement C2002).
This insulation is designed for hot and cold equipment and not to withstand load. For cold insulation factors are taken from table C2004.
The factors could deviate somewhat and are simplified compared to general industrial measuring methods.
Material: glass wool with aluminium sheeting.

Part	Insulation thickness in mm	Cost per m^2 thermal insulation
Flat parts, walls, channels and tank roofs	30	102.-
	60	113.-
	80	117.-
	100	126.-
	120	133.-
	140	140.-
	160	149.-
Vessels, tanks and heat exchangers	30	110.-
	60	120.-
	80	125.-
	100	136.-
	120	145.-
	140	152.-
	160	159.-
Part	**Equipment**	**Factor on m^2**
Spherical top, bottom or front.	Tanks/vessels/columns/heat exchangers	3,0 body part
Segment cap, ring cap, conical piece or point cap	Tanks/vessels/columns/heat exchangers	2,5 bodypart
Transition piece, conical piece, point cap	Tanks/vessels/columns/heat exchangers	2,0 body part
Flat head, cap, lid or end piece whether or not removable	Tanks/vessels/columns/heat exchangers	1,5 flat surface
Cabinets or doors	Turbines, pumps	2,0 flat surface

C2002 Method of measurement for equipment insulation

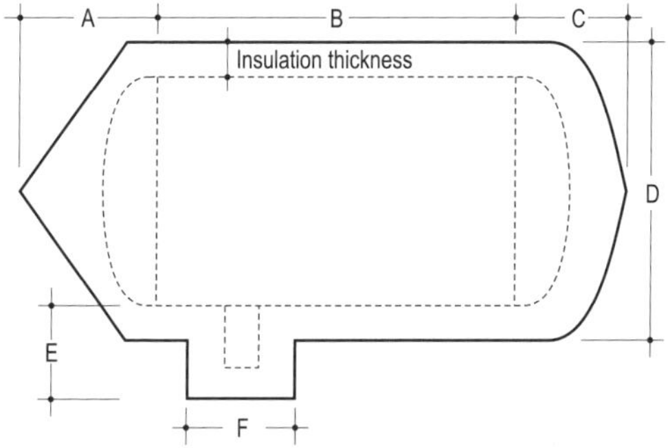

The parts of this example drawing are described in the table.

Part	Description
A	Height of a cap. Measured from tangent line to outside the insulation
B	Length of the body. Measured from tangent line to tangent line.
C	Height of spherical top (bottom, front). Measured from tangent line to outside the insulation
D	Diameter of the insulated equipment
E	Height of connection, nozzle, manhole, dome
F	Diameter of outside the insulation

C2 Insulation

C2003 Thermal insulation on piping

Costs are based on realized work, given per meter, and based on average built pipe systems with hot type insulation of glass wool with aluminium sheeting. The average price per meter is inclusive of all materials, labour as well as the insulation of components and valves.
The calculations are based on a piping system of 150 m. The costs of insulation on components and valves are added.
Finishing: glass wool with aluminium sheeting.
Pipe system of:
- 40 elbows;
- 10 tees;
- 10 valves;
- 150 m pipe (system length);
- 20 sets of flanges.

Nominal diameter in inches	Outside diameter in mm	Insulation thickness in mm	Price per m piping
1	33.4	50	47.-
2	60.3	50	55.-
3	88.9	50	62.-
4	114.3	50	70.-
6	168.3	50	85.-
8	219.1	50	104.-
		100	149.-
10	273	50	120.-
		100	177.-
12	323.8	50	136.-
		100	202.-
14	335.6	50	139.-
		100	222.-
16	406.4	50	163.-
		100	255.-
18	457.2	50	179.-
		100	280.-
20	508	50	193.-
		100	302.-
24	610	50	227.-
		100	349.-

C2 Insulation

C2004 Surcharges for cold insulation and sound

The surcharges are calculated by factors over m2 price on equipment and meter price on pipes. These can be used with the tables for heat insulation on equipment and pipes. The tables and measuring methods are approximations.
Cold insulation: PIR/PUR 50 mm with aluminium sheeting, armaflex 1,9 mm without additional finishing, foam glass 50 mm with aluminium sheeting.
Sound insulation: Rockwool, anti-drumming foil and aluminium sheeting.

Insulation type	Factor over cost per unit		
	PIR/PUR	Foam glass	Rock wool
Cold insulation	2.50	2.90	
Sound insulation			1.70

C2005 Examples insulation thickness

This table gives an indication of insulation thickness and can be used in conjunction with the tables for insulation on equipment and piping.

Insulation type	Temperature in °C	Insulation thickness in mm
Hot insulation	80	30
	130	60
	185	80
	245	100
	300	120
	360	140
	430	160
Cold insulation	0 till –20	30 till 60
	–20 till –50	60 till 140
	–50 till –70	140 till 160
Winterizing		30 till 60

C

C

D ELECTRICAL

D1 Energy supply

D1001 Electric power generation (rotating)

Diesel generator
Including:
- Installation;
- Connection.

Excluding:
- Building costs.

Variation
- Short time parallel operation: + € 4,000.-
- Long time parallel operation: + € 7,000.-

Generator	Capacity in kVA	Cost		
		Open set	Set in enclosure	Control Panel
Diesel generator	25	8,200.-	8,500.-	1,100.-
	60	11,200.-	12,800.-	1,100.-
	100	13,200.-	15,000.-	2,100.-
	150	21,400.-	20,300.-	2,100.-
	230	28,500.-	31,000.-	2,100.-
	275	34,600.-	38,300.-	3,100.-
	400	38,700.-	48,000.-	3,100.-
	550	48,900.-	60,800.-	4,300.-
	635	63,200.-	86,300.-	4,300.-
	730	80,500.-	91,600.-	5,300.-
	900	135,000.-		6,400.-
	1,000	181,000.-		6,400.-
	1,250	203,000.-		7,400.-
	1,350	214,000.-		9,600.-
	1,500	258,000.-		9,600.-
	1,750	301,000.-		21,300.-
	1,825	337,000.-		24,500.-
	2,250	347,000.-		24,500.-

D

D ELECTRICAL

D1 Energy supply

D1002 Emergency power supply (static) 3 x 400 V / 230 V

UPS-emergency power supply in case of power failure.
For industrial facilities.
Input: 3 x 400 V, 50 Hz.
Output: 230 V, 50 Hz.
Emergency power supply by industrial NiCad batteries.

Capacity in KVA	Autonomy time in min.	Cost per unit
5	30	30,800.-
	60	32,900.-
	180	49,600.-
	300	70,000.-
10	30	38,500.-
	60	46,300.-
	180	77,900.-
	300	109,000.-
15	30	46,100.-
	60	59,400.-
	180	102,000.-
	300	173,000.-
20	30	60,600.-
	60	74,700.-
	180	134,000.-
	300	191,000.-
30	30	77,900.-
	60	95,400.-
	180	194,000.-
	300	302,000.-
40	30	101,000.-
	60	124,000.-
	180	249,000.-
	300	381,000.-
50	30	109,000.-
	60	144,000.-
	180	254,000.-
	300	458,000.-
60	30	134,000.-
	60	162,000.-
	180	335,000.-
	300	588,000.-
75	30	152,000.-
	60	191,000.-
	180	416,000.-
	300	735,000.-

D ELECTRICAL

D1 Energy supply

D1003 Emergency power supply (static) 3 x 400 V

UPS-emergency power supply in case of power failure.
For industrial facilities.
Input: 3 x 400 V, 50 Hz.
Output: 3 x 400 V, 50 Hz.
Emergency power supply by industrial NiCad batteries.

Capacity in KVA	Autonomy time in min.	Cost per unit
10	30	48,700.-
	60	54,100.-
	180	85,500.-
	300	121,000.-
15	30	55,200.-
	60	69,300.-
	180	113,000.-
	300	162,000.-
20	30	69,300.-
	60	77,900.-
	180	126,000.-
	300	144,000.-
30	30	83,200.-
	60	98,400.-
	180	200,000.-
	300	305,000.-
50	30	109,000.-
	60	145,000.-
	180	312,000.-
	300	366,000.-
75	30	156,000.-
	60	199,000.-
	180	419,000.-
	300	742,000.-
100	30	191,000.-
	60	256,000.-
	180	524,000.-
	300	863,000.-
150	30	269,000.-
	60	351,000.-
	180	752,000.-
	300	1,213,000.-

D1004 Emergency power supply (static) 230 - 240 V

UPS-emergency power supply in case of power failure.
Input: 230 - 240 V, 50 Hz.
Output: 230 - 240 V, 50 Hz.
Emergency power supply by batteries.

Capacity in KVA	Autonomy time in min.	Cost per unit
3	10	5,700.-
	25	7,700.-
5	10	7,300.-
	25	9,200.-
10	10	9,700.-
	25	11,600.-

D ELECTRICAL

D1 Energy supply

D1005 Emergency power supply (static) 3 x 400 - 415 V

UPS-emergency power supply in case of power failure.
For computers and PLC-control.
Input: 3 x 400 - 415 V, 50 Hz.
Output: 3 x 400 - 415 V, 50 Hz.
Emergency power supply by batteries.

Capacity in KVA	Autonomy time in min.	Cost per unit
10	10	24,200.-
	30	24,900.-
	60	26,200.-
15	10	26,200.-
	30	27,800.-
	60	31,100.-
20	10	29,000.-
	30	31,600.-
	60	33,400.-
30	10	32,900.-
	30	36,200.-
	60	43,800.-
40	10	41,000.-
	30	48,100.-
	60	52,400.-
60	10	45,300.-
	30	52,900.-
	60	62,200.-
80	10	56,300.-
	30	68,100.-
	60	72,600.-
100	10	62,200.-
	30	74,200.-
	60	89,000.-
120	10	85,300.-
	30	99,000.-
	60	116,000.-
160	10	102,000.-
	30	119,000.-
	60	142,000.-
220	10	128,000.-
	30	150,000.-
	60	189,000.-

D

D1 Energy supply

D1006 High and low voltage underground electrical power cables

High voltage cable VG-YLK Dca - 10 kV - Cu.
Low voltage braided cable, YMvKas Dca 750 V - Cu.
3- or 4-core.
Including:
- Supply and install cable in the ground.
Excluding:
- Excavation and paving.

Cable	Number of cores	Core in mm^2	Cost each
High Voltage cable (VG-YLK)	3	25	57.-
		35	77.-
		50	92.-
		70	110.-
		95	124.-
		120-	139.-
		150	168.-
		185	187.-
Low Voltage cable	3	2.5	7.-
		4	8.-
		6	11.-
		10	14.-
		16	22.-
		25	29.-
		35	35.-
		50	46.-
		70	63.-
		95	80.-
		120	106.-
		150	125.-
		185	160.-
		240	211.-
	4	2.5	7.-
		4	8.-
		6	12.-
		10	18.-
		16	26.-
		25	34.-
		35	43.-
		50	56.-
		70	71.-
		95	110.-
		120	136.-
		150	173.-
		185	207.-
		240	260.-

D1007 Price factor high and low voltage electrical power cables CPR certified B2ca/Cca

Cost factor for CPR certified cable	
Cable	**Cost factor**
B2ca	1.17
Cca	1.13

D

D1 Energy supply

D1008 Cable connections and joints for high and low voltage underground cables

High voltage cable VG-YLK Dca - 10 kV - Cu.
Low voltage braided screen cable, VMvKas or YMvKas Dca 750 V - Cu.
3 or 4 core.
Including:
- Connection materials;
- Installation.

Cable joint	Type	Core in mm^2	Cost each
High Voltage cable	Cable end	10 - 70	990.-
		95 - 150	1,190.-
		185 - 240	1,520.-
	Joint	10 - 70	1,360.-
		95 - 150	1,680.-
		185 - 240	2,180.-
Low Voltage cable	Cable end	2,5 - 6	70.-
		10 - 35	130.-
		50 - 95	200.-
		120 - 240	340.-
	Joint	2,5 - 6	210.-
		10 - 35	310.-
		50 - 95	380.-
		120 - 240	690.-

D1009 Low voltage signal underground cables

Low voltage braided screen control and signal multicore cables YMvKas Dca.
Including:
- Supply and install cable in the ground.

Excluding:
- Excavation and paving.

Number of cores	Core in mm^2	Cost per m
7	1.5	8.-
	2.5	12.-
12	2.5	19.-
19	1.5	21.-
	2.5	28.-
24	2.5	37.-

D1010 Price factor low voltage signal cables CPR certified B2ca/Cca

Cost factor for CPR certified cable	
Cable	**Cost factor**
B2ca	1.19
Cca	1.23

D ELECTRICAL

D1 Energy supply

D1011 Cable ends and joints for low voltage underground cables

Cable ends and joints for braided screen control and signal multicore cables YMvKas.
Including:
- Connection materials;
- Installation.

Cable connection	Number of cores	Cost each
Cable end	7	84.-
	12	130.-
	19	163.-
	24	202.-
Joint	7	213.-
	12	250.-
	19	378.-
	24	414.-

D1012 Prefabricated transformer station

Concrete prefabricated compact station.
Oil cooled transformer 10 / 0.4 kV - 630 kVA.
Switchboard 10 kV - 250 MVA with 3 load isolators.
Switchboard 0.4 kV with:
- 1,000 A circuit breaker;
- 3 outgoing feeders 400 A;
- 3 outgoing feeders 250 A;
- 2 spare positions for outgoing feeders.

Including:
- Supply;
- Installation;
- Connection;
- Lighting;
- Internal grounding;
- Gratings;
- Painting.

Excluding:
- Foundation;
- Building permits, administrative expenses, etc.;
- Connection of outgoing low voltage cables;
- All incoming and outgoing cables.

Type	Cost each
Station without measurement in a private network	47,300.-
Station with an impure measurement in a private network	48,500.-
Station with an impure measurement in an authority network	39,800.-
Station with a pure measurement in an authority network	38,900.-

D

D1 Energy supply

D1013 High voltage switch installation

Installed and connected.
Single rail system.
Circuit breaker with protection.
Load breaker switch with load breaker switches and safety fuses, cast in artificial resin. The system consists of two cable fields and one transformer field.
Excluding:
- Architectural facilities.

Variation
The price of a double rail system is depending on the specification.
This can be 40 - 80% higher than a single rail system.

Type	Operating voltage in kV	Short circuit power in MVA	Type of feeder	Unit	Cost per unit	
					From	Till
Circuit breaker	6	150	Incoming feeder unit	Up to 630 A	12,400.-	22,300.-
			Outgoing feeder unit	Up to 630 A	11,700.-	19,400.-
	10	250	Incoming feeder unit	Up to 630 A	15,000.-	21,800.-
			Outgoing feeder unit	Up to 630 A	12,900.-	18,900.-
	10	350	Incoming feeder unit	Up to 630 A	18,800.-	25,000.-
			Outgoing feeder unit	Up to 630 A	15,000.-	20,400.-
	10	500	Incoming feeder unit	Up to 630 A	22,300.-	38,800.-
			Outgoing feeder unit	Up to 630 A	16,100.-	25,000.-
	20	800	Incoming feeder unit	Up to 630 A	28,300.-	52,100.-
			Outgoing feeder unit	Up to 630 A	24,500.-	45,400.-
	25 to 30	1,500	Incoming feeder unit	Up to 630 A	38,800.-	60,300.-
			Outgoing feeder unit	Up to 630 A	29,400.-	53,100.-
Load switch	10	250		630 A	12,400.-	17,800.-

D1014 Standard transformers, 10 kV to 400 V, oil cooled

Oil cooled transformers for indoor installation.
Installed and connected.
Capacity: 10 kV to 400 V.
Including:
- Conservator;
- Gas relay or nitrogen blanket.

Excluding:
- Architectural facilities;
- Discounts;
- Cable work.

Variation
- Quantity price reduction is possible;
- If installed indoors, a reduction of € 2,500.- is applicable.

Capacity in kVA	Cost each	
	From	Till
100	6,500.-	7,900.-
250	7,900.-	9,300.-
400	9,200.-	10,500.-
630	11,700.-	13,100.-
1,000	14,300.-	15,800.-
1,250	17,000.-	18,400.-
1,600	19,600.-	21,000.-
2,000	22,200.-	23,700.-
2,500	24,800.-	26,300.-
3,150	32,700.-	39,400.-

D ELECTRICAL

D1 Energy supply

D1015 Standard transformers, 10 kV to 400 V, cast resin dry type

Cast resin dry type transformers for indoor installation.
Installed and connected.
Capacity: 10 kV to 400 V.
Including:
- Temperature monitoring.

Excluding:
- Architectural facilities;
- Discounts;
- Cable work.

Variation
- Quantity price reduction is possible;
- The price of a transformer 20 kV to 400 V is 20% higher.

Capacity in kVA	Cost each	
	From	Till
100	6,500.-	7,900.-
250	9,100.-	10,500.-
400	11,700.-	13,100.-
630	14,300.-	15,700.-
1,000	16,900.-	18,300.-
1,250	19,500.-	21,000.-
1,600	22,200.-	23,600.-

D

D1016 Standard transformers, 20 - 30 kV to 10 kV

Transformers for outdoor installation.
Capacity: 20 - 30 kV to 10 kV.
Including:
- Switches;
- Junction boxes;
- Insulators;
- Cable connections.

Excluding:
- Architectural facilities;
- Discounts;
- Cable work.

Capacity in MVA	Cost each	
	From	Till
1.25	30,600.-	41,900.-
1.6	35,100.-	44,500.-
2	41,600.-	52,400.-
2.5	47,500.-	58,900.-
3	52,700.-	65,500.-
4	62,500.-	77,900.-
5	74,800.-	92,200.-
6	88,500.-	109,900.-
8	113,000.-	140,000.-
10	139,000.-	171,000.-
12	161,000.-	200,000.-
15	212,000.-	243,000.-

D ELECTRICAL

D1 Energy supply

D1017 Standard transformers, 20 kV to 400 V

Transformers for outdoor installation.
Installed and connected.
Capacity: 20 kV to 400 V.
Including:
- Switches;
- Junction boxes.

Excluding:
- Architectural facilities;
- Discounts;
- Cable work.

Capacity in MVA	Cost each	
	From	Till
1.25	23,500.-	32,700.-
1.6	26,100.-	38,000.-
2	31,300.-	43,200.-
2.5	40,400.-	49,100.-
3	45,600.-	55,000.-

D

D1018 Low voltage switch in enclosure

Type
Short-circuit capacity: 60 kA.
In busbar: 2,000 A.
Nominal voltage: 400 V.
Incoming feeder: Switch disconnector.
Outgoing feeder: Safety fuse.
Plate steel and compartmented.
Including:
- Assembly and start-up.

Excluding:
- Architectural facilities;
- Discounts;
- Connections.

Type of power field	Cost	
	From	Till
Incoming feeder 1,600 A	11,400.-	13,500.-
10 outgoing feeders 250 A	11,400.-	13,500.-
Incoming feeder 1,600 A	11,400.-	13,500.-
10 outgoing feeders 400 A	16,600.-	22,900.-
Incoming feeder 1,600 A	11,400.-	13,500.-
10 outgoing feeders 630 A	56,200.-	67,600.-

D2 Energy conversion

D2001 Electrical power installation

Based on approximately 30 users, small as well as greater motors with an average rate between 18.5 kW and 22 kW.

Type

Distribution boards:

- Power distribution board; synthetic enclosed distribution boxes and motor protection switches at the motors.

Central motor control cabinet:

- Power distribution board; central motor control cabinet with contactors and control relays.

Type	Cost each
Distribution board	570.-
Central motor control cabinet	740.-

D2002 Motor control centre (MCC)

The following subdivision is used to determine the total price:

- < 4 kW: 60%;
- 4 - 18.5 kW: 30%;
- > 18.5 kW: 10%.

Variation

Prices strongly depend on the MCC specification, complexity of the circuit and the ratio between the amount of groups and the total of the installed capacity.

Installed motor power in kW	Cost each
15	2,800.-
35	4,000.-
70	5,700.-
100	6,800.-
150	7,900.-
200	9,200.-
250	10,200.-
350	11,400.-
450	13,500.-
550	15,600.-
600	18,700.-

D2003 Motor control switch (MCS)

Completely installed and connected.
Closed enclosure, IP54.
Switch-on method: Direct (DOL).

Motor power in kW	Cost each
To 5.5	2,900.-
From 7.5 to 11	3,400.-
From 15 to 18.5	4,600.-
From 22 to 30	5,000.-
From 37 to 55	6,000.-
From 75 to 90	7,000.-

D2004 Low voltage power cables YMvK Dca in existing cable trays or conduits

Supply and install (both vertical and horizontal).

Number of cores	Core in mm^2	Cost each
3	2.5	7.-
	4	8.-
	6	11.-
	10	13.-
	16	17.-
	25	24.-
	35	31.-
	50	47.-
	70	62.-
	95	80.-
	120	103.-
	150	125.-
	185	155.-
	240	209.-
4	2.5	7.-
	4.	8.-
	6	12.-
	10	16.-
	16	22.-
	25	30.-
	35	36.-
	50	54.-
	70	72.-
	95	92.-
	120	118.-
	150	145.-
	185	178.-
	240	240.-
5	2.5	10.-
	4	11.-
	6	13.-
	10	18.-
	16	25.-
	25	36.-
	35	47.-
	50	71.-
	70	93.-
	95	121.-
	120	154.-

D2005 Price factor Low voltage power cables in existing cable trays or conduits CPR certified Cca/B2ca

Cost factor for CPR certified cable

Cable	Cost factor
Cca	1.13
B2ca	1.17

D

D2006 Cables for control and signals, YMvK Dca

Supply and install (vertical or horizontal).

Location	Number of cores	Core in mm^2	Cost each
In existing horizontal cable tray	6	1.5	9.-
		2.5	11.-
	7	1.5	9.-
		2.5	11.-
	8	1.5	9.-
		2.5	11.-
	12	1.5	11.-
		2.5	13.-
	19	1.5	13.-
		2.5	16.-
	24	1.5	15.-
		2.5	20.-
	30	1.5	21.-
		2.5	25.-
	37	1.5	19.-
		2.5	28.-
In vertical cable tray	6	1.5	13.-
		2.5	14.-
	7	1.5	14.-
		2.5	15.-
	8	1.5	14.-
		2.5	15.-
	12	1.5	15.-
		2.5	18.-
	19	1.5	18.-
		2.5	21.-
	24	1.5	20.-
		2.5	24.-
	30	1.5	25.-
		2.5	29.-
	37	1.5	24.-
		2.5	34.-

D2007 Price factor Cables for control and signals CPR certified Cca/B2ca

Cost factor for CPR certified cable	
Cable	Cost factor
Cca	1.19
B2ca	1.23

D2008 Cable junction box for cables YMvK (signal and control)

Junction box for cables YMvK with a core diameter of 1.5 mm^2 or 2.5 mm^2 per core.
Class: Standard, EEx e or EEx i.
Enclosure: Plastic or Stainless steel.

Number of cores	Cost each	
	Plastic	Stainles steel
6	140.-	430.-
7	140.-	430.-
8	170.-	630.-
12	230.-	650.-
19	270.-	710.-
24	330.-	740.-
30	360.-	770.-
37	400.-	980.-

D2 Energy conversion

D2009 Low voltage motor, EEx n

Type: Horizontal baseplate construction B3 (230 - 690 V, 50 Hz, 3 phases).
Speed: 3,000 rpm (2 poles).
Class: Standard, EEx n.
Protection class: IP55, fully closed.
Insulation class: F.
Excluding:
- Installation;
- Connection.

Variation
The DIN 42673 construction B5 (horizontal flanged mounted) or V1 (vertical flanged mounted) motors are 5 - 12 % more expensive than horizontal baseplate construction motors B3. This factor is inversely proportional to the motor power.

Motor power in kW	Cost each
0.37	220.-
0.55	230.-
0.75	250.-
1.1	310.-
1.5	340.-
2.2	430.-
3	540.-
4	660.-
5.5	830.-
7.5	1,000.-
11	1,300.-
15	1,700.-
18.5	2,200.-
22	2,400.-
30	3,200.-
37	4,000.-
45	4,800.-
55	5,800.-
75	7,900.-
90	9,500.-
110	11,800.-
132	14,200.-

D

D2010 Price factor rpm/electrical power for low voltage motor EEx n

Cost factor for speed rate			
Poles	rpm	Capacity in kW	Cost factor
4	1,500	≤ 30	1
		> 30	1
6	1,000	≤ 30	1.2
		> 30	1.2
8	750	≤ 30	1.8
		> 30	1.4

D2011 Low voltage motor, EEx e

Type: Horizontal baseplate construction B3 (230 - 690 V, 50 Hz, 3 phases).
Speed: 3,000 rpm (2 poles).
Class: Standard, EEx e.
Protection class: IP55, fully closed.
Temperature class: T3 (200 °C).
Insulation class: F.
Excluding:
- Installation;
- Connection.

Variation
The DIN 42673 construction B5 (horizontal flanged mounted) or V1 (vertical flanged mounted) motors are 5 - 12 % more expensive than horizontal baseplate construction motors B3. This factor is inversely proportional to the motor power.

Power rate in kW	Cost each
0.37	230.-
0.55	250.-
0.75	260.-
1.1	320.-
1.5	430.-
2.2	540.-
3	610.-
4	800.-
5.5	940.-
7.5	1,300.-
11	2,000.-
15	2,400.-
18.5	3,000.-
22	3,600.-
30	4,400.-
37	5,800.-
45	8,300.-
55	11,700.-

D2012 Price factor rpm/electrical power for low voltage motor EEx e

Cost factor for speed rate			
Poles	rpm	Capacity in kW	Cost factor
4	1,500	≤ 30	1
		> 30	1
6	1,000	≤ 30	1.2
		> 30	1.2
8	750	≤ 30	1.8
		> 30	1.4

D2 Energy conversion

D2013 Low voltage motor, EEx d

Type: Horizontal baseplate construction B3 (230 - 690 V, 50 Hz, 3 phases).
Speed: 3,000 rpm (2 poles).
Class: standard, EEx d.
Protection class: IP55, fully closed.
Temperature class: T1 - T4.
Insulation class: F.
Excluding:
- Installation;
- Connection.

Variation
The DIN 42673 construction B5 (horizontal flanged mounted) or V1 (vertical flanged mounted) motors are 5 - 12 % more expensive than horizontal baseplate construction motors B3. This factor is inversely proportional to the motor power.

Power rate in kW	Cost each
0.37	570.-
0.55	670.-
0.75	760.-
1.1	820.-
1.5	960.-
2.2	1,000.-
3	1,100.-
4	1,400.-
5.5	1,700.-
7.5	2,200.-
11	2,600.-
15	3,300.-
18.5	4,000.-
22	4,700.-
30	6,100.-
37	7,200.-
45	8,300.-
55	11,100.-
75	14,400.-
90	17,500.-
110	21,700.-
132	25,900.-

D

D2014 Price factor for low voltage motor EEx d

Cost factor for speed rate			
Poles	rpm	Capacity in kW	Cost factor
4	1,500	≤ 30	1
		> 30	1
6	1,000	≤ 30	1.2
		> 30	1.2
8	750	≤ 30	1.8
		> 30	1.4

D2 Energy conversion

D2015 Connecting low voltage motors

Type: Horizontal baseplate construction B3 (230 - 690 V, 50 Hz, 3 phases).
Connection of low voltage motors, directly online, 2 sided, excluding cables.

Power rate in kW	Cost each
Up to 18.5	420.-
From 22 to 37	660.-
From 45 to 110	840.-

D2016 Temperature class low voltage motors

Maximum surface temperature at 40 °C ambient temperature.

Temperature class	Cost each
T1	520.-
T2	340.-
T3	230.-
T4	160.-
T5	110.-
T6	90.-

D2017 Variable frequency converters

Type: 400 V, 50 Hz, wall mounted.
Protection class: IP 21.
Including:
- Installation.

Power rate in KW	Cost each
3	1,700.-
7.5	2,500.-
18.5	5,100.-
30	6,500.-
45	9,000.-
75	12,700.-
110	16,500.-
132	20,900.-

D2 Energy conversion

D2018 Sockets outlets

CEE-type: DIN VDE 0623, EN 60309-2.
Including:
- Installation.

Excluding:
- Cable.

Type	Model	Numbers of Poles	Current in A	Voltage in V	Hour position	Cost each
Wall mounted	IP44	2	16	42	12H	230.-
		3	16	230	6H	180.-
			32	230	6H	230.-
			63	230	6H	300.-
		4	16	400	6H	180.-
		5	16	400	6H	180.-
			32	400	6H	190.-
	IP67	3	16	230	6H	240.-
		4	125	400	6H	650.-
		5	16	400	6H	250.-
			32	400	6H	270.-
Wall mounted disconnectable	IP44	5	16	400	6H	540.-
			32	400	6H	580.-
	IP67	4	16	400	6H	550.-
		5	16	400	6H	580.-
			32	400	6H	640.-
			63	400	6H	710.-
			125	400	6H	830.-
Panel mounted	IP44	2	16	42	12H	240.-
		3	16	230	6H	240.-
			32	230	6H	250.-
		4	16	400	6H	240.-
			32	400	6H	250.-
		5	16	400	6H	240.-
			32	400	6H	260.-
			63	400	6H	350.-
	IP67	3	16	230	6H	260.-
		4	16	400	6H	260.-
		5	16	400	6H	270.-
			32	400	6H	270.-
			63	400	6H	360.-
			125	400	6H	710.-
Apparatus inlet socket	IP44	2	16	42	12H	260.-
		3	32	230	6H	260.-
		4	16	400	6H	250.-
			32	400	6H	260.-
		5	16	400	6H	260.-
			32	400	6H	270.-

D

D2 Energy conversion

D2019 Lighting fixtures

Fixtures for fluorescent tubes and floodlights.
Type: According to IEC 144.
The explosion proof construction (NEN 3125):
- If 'E' is noted at the column 'model' a construction with increased safety (Ex e G3) is intended.

Including:
- Installation.

Excluding:
- Scaffolding;
- Cables.

Type fixture	Specification	Model	Maximum power	Cost each
Fluorescent tubes fixture	Striplight	IP 21	2 x TL 36 W	240.-
	Dripproof	IP 54	2 x TL 36 W	290.-
	Waterproof	IP 65	2 x TL 36 W	350.-
	Explosion proof	E'	1 x TL 18 W	480.-
			1 x TL 36 W	500.-
			2 x TL 36 W	550.-
	Pole mounted	IP 54	2 x TL 18 W	290.-
			2 x TL 36 W	290.-
	Pole mounted - explosion proof	E'	2 x TL 18 W	540.-
			2 x TL 36 W	560.-
Sodium lamp fixture	Pole mounted	IP 54/53	150 W	460.-
Mercury lamp fixture	Pole mounted	IP 54/53	125 W	380.-
Floodlight	10,000 cd	IP 55	1,000 W	860.-
	1,000,000 cd	IP 55	2,000 W	1,100.-

D

D2020 Lighting poles

Galvanized.
Completely installed.
Excluding:
- Lighting fixture.

Height fixture in m	Shape	Cost each
3	Cylindrical	590.-
5	Cylindrical	650.-
8	Cylindrical	910.-
10 till 12	Cylindrical	1,600.-
25	Lattice lighting pole	25,300.-

D2021 Electrical heat tracing

Electrical heat tracing, such as:
- Serial heating cables (mineral or plastic insulated);
- Self-regulating and/or power limited (parallel) cables.

Including:
- Termination kits, (warning) labels, small junction boxes;
- Installation onto preinstalled piping, tubing, instruments, etc.

Temperature to be maintained in °C	Maximum temperature exposure of cable in °C	Cost each per m pipe
5	85	69.-
65	85	82.-
120	190	98.-
215	260	111.-

D2 Energy conversion

D2022 Installation of fibre optic cable

Complete installation of HDPE conduit.
Maximal length 125 m.
Including:
- Inserting the fibre optic cable.

Number of fibres	Cost each
12	16.-
24	20.-
36	22.-
48	25.-

D2023 Fibre optic interconnecting cable joint

Complete joint sleeve.

Number of fibres	Cost each
12	3,100.-
24	3,500.-
36	4,100.-
48	4,800.-

D2024 Fibre optic cable, cable end connection

Complete finishing the fibre optic cable on a patch panel.
Including:
- Measurement / testing on 2 wave length, 2-side.

Number of fibres	Cost each
12	1,900.-
24	3,900.-
36	5,600.-
48	7,500.-

E INSTRUMENTATION AND PROCESS CONTROL

E0 Explanation of instrumentation and process control

E0001 General remarks

Design:
- Cast steel according ASTM A-216-WCC or WCB. Or equal to DIN;
- Stainless steel according ASTM A351-CF8M or CF8C. Or equal to DIN.

Without special notes the standard will be:
- Non Ex model and suitable for EEx(i) application;
- Sealing minimal IP65 according IEC/EN-60529;
- Transmitters signal output 4 -20 mA with HART protocol.

To compare pressure traps from PN into lbs ANSI-B use:
- PN 10 - 150 lbs;
- PN 16 - 150 lbs;
- PN 40 - 300 lbs;
- PN 100 - 600 lbs.

To compare pipe diameters DIN – ANSI use:
- DN15 - 0.5'';
- DN20 - 0.75'';
- DN25 - 1'';
- DN40 - 1.5'';
- DN50 - 2'';
- DN80 - 3'';
- DN100 - 4'';
- DN150 - 6'';
- DN200 - 8'';
- DN250 - 10''.

Variation

Quantum discount is possible. Special conditions like material and/or test certification are not included.

Abbreviations	Description	Remark
SS	Stainless steel	Quality AISI 304 of 316
DCS	Distributed Control System	
DN	Nominal diameter in mm	According DIN 2401 part 1
PN	Nominal pressure	According DIN 2402
lbs	Nominal pressure	According ANSI-B 16.5
inch	Nominal diameter in mm	According ANSI-B 16.5
SCV	Short cone venturi	
NPT	Screw thread according American standard	
PTFE	Polytetrafluorethylene (Teflon/Fluon)	
EEx i	Intrinsic safe design	According IEC 60079-11
EEx d	Explosive safe application, compression resistant	According IEC 60079-1
WD	Water resistant IP 55	According IEC 60529
PLC	Programmable Logic Controller	
PVC	Poly vinyl chloride	
TC	Thermo couple	
Pt100	Electrical resistant	

E

E INSTRUMENTATION AND PROCESS CONTROL

E1 Systems, in general

E1001 Decentral digital control system

Units for blind control room: unit system, usually in a blind control room, connected with field cables.
Operator interface: situated in the control room where several blind control room, systems could be connected to.
Minimal 2 operating stations to calculate per separate process unit.

Part	Specification	Cost each
Unit for blind control room	Unit for 8 arranged ports	56,000.-
	Allowance for intrinsic safe input or output	870.-
	Allowance for a redundant model for a maximum of 64 outputs	117,000.-
	Allowance for a redundant model for 64 additional outputs	117,000.-
	Unit for analogue and discrete signals for data-acquisition	19,900.-
	Unit for analogue and discrete signals for mV signals	21,600.-
	Allowance per in- and output	480.-
	Allowance for intrinsic safe in- or output	800.-
Operator interface	Operating position, with screen, keyboard and data bus	79,400.-
	Allowance for printer	5,600.-
	Allowance for video-copier	32,400.-
	Allowance for console with four line writers	40,800.-

E1002 Free Programmable Control system (PLC)

Variation
Price of a central unit in a redundant application will be 2.5 times more expensive.

Part	Specification	Amount of discrete in- and outputs	Cost each
Central processing unit		256	27,000.-
		1	37,000.-
		2	50,000.-
Panel with mounting frame and wiring			45,000.-
Programming unit	With printer		22,000.-
Allowance	Digital input		85.-
	Analogue input		780.-
	Digital output		170.-
	Analogue output		960.-
	Intrinsic safety digital in- or output		540.-

E

E INSTRUMENTATION AND PROCESS CONTROL

E1 Systems, in general

E1003 Installation of instruments

The cost of a total installed control loop (instrument loop) depends on:
- Selection of instruments and material;
- Infra structure and signal transfer (conventional cables, fieldbus, radio or phone connection);
- Complexity of the control action (e.g direct reading, alarm, feed forward control);
- Safety classification (SIL, ATEX, PED).

Including:
- Instruments (e.g. sensors, signal transmitter, control valves, valves);
- All required material for connecting to process and instrument air;
- Cables, cable trays and additional material;
- Hardware and software for process control system;
- Calibrating, testing and put into operation;
- Installation and program costs.

Variation
If installing during operational production the costs will increase 15 - 20%.

Description	Cost each
Single measuring	68,000.-
Complex control circuit	135,000.-

E1004 Panels for process control

Cost of panels, wires and/or pneumatic tubes installed by the supplier.
Including:
- Internal wires;
- Installation materials.

Excluding:
- Instruments.

Description	Specification	Unit	Cost per unit	
			From	Till
Straight panels	Maximum 2.5 m height	m width	9,100.-	13,800.-
Desk panels	Maximum 1.8 m height	m width	7,900.-	19,700.-
Local panels IP 55	Maximum 2 m height	m width	14,700.-	32,200.-
Semigraphic panels	Dimensions 300 x 300 mm	each		700.-
	With 100 LEDs, completely wired	each		3,900.-

E

E INSTRUMENTATION AND PROCESS CONTROL

E2 Instrumentation for quality measurement

E2001 Analysis apparatus for oil refinery

Measurement	According	Cost each	
		Analyser	Sampling rack
Boiling point	ASTM D-86	49,000.-	23,000.-
Cloud-, Freeze- of Pour-point	ASTM D-97	60,000.-	24,000.-
Flash-point by thermo couple detection	ASTM D-56	66,000.-	27,000.-
Vapour pressure	ASTM D-323	60,000.-	24,000.-
Colour analysis	ASTM D-1500	23,000.-	17,000.-
Viscosity by differential pressure measurement by capillary		60,000.-	24,000.-

E2002 Sulphur analysis apparatus

Measurement	Specification	Cost each	
		Analyser	Sampling rack
H_2S/SO_2 - ratio	Photometric ultraviolet absorption	79,000.-	17,500.-
Sulphur liquid	Radiation absorption	73,000.-	16,200.-
H_2S for personal protection	With sensitive tape, sample system, pump flow meter	39,000.-	17,500.-
Sulphur gas	Measuring by combustion to SO_2-infrared	52,000.-	8,300.-
	H_2S in a process stream	32,000.-	18,600.-

E2003 Gas chromatography

Additional surcharge for flame ionization detection system:
- Maximum of 4 components per measured current.

Including:
- Program-unit.

Measurement	Specification	Cost each	
		Analyser	Sample rack
Thermal conduction detection	For one stream	61,000.-	18,300.-
	For two streams	62,000.-	24,100.-
	For four streams	65,000.-	39,100.-
Allowance for flame ionization detection system		5,000.-	
Maintenance recorder		8,000.-	
Calibrating- and propellant per cylinder			700.-

E2004 General components analysing instruments for gas

Measurement	Specification	Cost each	
		Analyser	Sampling rack
Infrared absorption		30,000.-	17,500.-
Ultraviolet	Single ray (double wave length)	20,000.-	12,100.-
	Double ray	39,000.-	14,500.-
Thermal conduction detection		106,000.-	8,400.-

E2 Instrumentation for quality measurement

E2005 Chemical and physical measuring

PH measuring, ion-selective measuring and conductivity measuring.
Including:
- SS flow fittings;
- Electrode.

Ion concentrate analyser including:
- Chemical reactant.

Measurement principle	Cost each	
	Analyser	Sampling rack
pH-measurement	5,400.-	3,900.-
Ion-selective measurement	6,200.-	5,300.-
Conduction measurement	4,000.-	5,000.-
Ion-concentrate analysis	26,900.-	5,000.-

E2006 Density measuring

Measurement principle	Specification	Cost each	
		Analyser	Sampling rack
Fluid in-feed type	Frequency damping (mA output)	15,700.-	8,100.-
	Gamma radiation absorption	22,900.-	12,100.-
	U-tube with frequency converter	12,600.-	8,100.-
Gas in-feed type	Frequency damping (mA output)	22,400.-	12,100.-

E2007 Waste gas analysis equipment

Measuring unit	Specification	Cost each	
		Analyser	Sampling rack
Smoke density	Photoelectric	8,800.-	
Oxygen content	Infeed type zirconium oxide cell, L 1,000 mm	19,300.-	
	Paramagnetic	14,300.-	9,800.-
Sulphur dioxide		66,500.-	12,200.-
	Chemical-fysical, electrical conduction in reaction fluid	21,100.-	10,100.-

E2008 Moisture measuring

Location	Measurement principle	Cost each	
		Analyser	Sampling rack
Gas	Difference in crystal oscillation between sample- and reference gas	42,400.-	8,900.-
	Infrared absorption	22,900.-	8,900.-
Dew point/hygrometer	Aluminiumoxide detection	8,700.-	6,900.-
	Lithiumchloride-/rhodium cell	5,300.-	11,000.-
Water in fluid	Titration automatic	56,000.-	14,400.-

E2 Instrumentation for quality measurement

E2009 Quality measuring for boiler feed water and waste water

Measuring equipment based on pressure difference.
Prices are inclusive difference pressure transmitters, process valves, and required connecting pipe and fittings.

Measuring unit	Measuring principle	Cost each	
		Analyser	Sampling rack
COD-analyser	Photometric	23,100.-	8,400.-
Turbidity	Ultraviolet absorption	21,200.-	8,400.-
Total hydrocarbons		49,600.-	8,400.-
Sodium trace		102,600.-	9,500.-
Dissolved oxygen		6,800.-	5,700.-
Oil in water	Photometric (floating oil detection)	23,100.-	8,400.-
Oil in boiler feed water		30,400.-	6,600.-

E

E3 Instrumentation for volume, flow and weight

E3001 Orifice plate

Material: SS.
Specification:
- Welded textplate, thickness 3 mm;
- Pressure class up to 600 lbs.

Calculate orifice plate: € 60.-.

Diameter in inches	Thickness in mm	Cost each
1	5	120.-
2	5	140.-
3	5	160.-
4	5	230.-
6	5	310.-
8	8	610.-
10	8	1,150.-

E3002 Orifice flanges (measuring flanges without valves)

Material: Carbon steel.
Specification:
- Set of 2 welding neck flanges each with one connection point;
- Pressure class up to 300 lbs.

Variation
Stainless steel: + 25% up to + 50%.

Diameter in inches	Cost each
1	370.-
2	380.-
3	510.-
4	640.-
6	850.-
8	1,290.-
10	1,920.-

E

E3003 Venturi

Material: Carbon steel.
Specification:
- Including welded neck flanges;
- Two connecting points for differential pressure measuring, 1/2" rigde piece;
- Pressure class up to 300 lbs.

Diameter in inches	Cost each
3	3,700.-
4	4,400.-
6	6,200.-
8	9,000.-
10	10,900.-

E3 Instrumentation for volume, flow and weight

E3004 Pitot tube (annubar)

Material: Stainless steel.
Specification:
- Plug-in tube;
- Flanged process connection 300 lbs / 1,5'';
- Two connecting points for differential pressure measuring, 1/2'' weld connection.

Pressure class in lbs	Tube diameter in inches	Cost each
Till 300 lbs	3	1,650.-
	4	1,650.-
	6	1,650.-
	8	1,650.-
	10	2,070.-
	12	2,070.-
	14	2,070.-
	16	2,480.-
Till 600 lbs	3	1,650.-
	4	1,650.-
	6	1,650.-
	8	1,650.-
	10	2,070.-
	12	2,070.-
	14	2,070.-
	16	2,480.-

E3005 Flow gauge

Material: Carbon steel casing including safety glass and stainless steel internal mechanism.
Specification:
- Flanged process connection;
- Equipped with internal rotating mechanism;
- Pressure class up to 300 lbs.

Tube diameter in inches	Cost each
1	290.-
2	420.-
3	710.-
4	990.-
6	1,690.-

E3006 Purge flow meter (Variable area flow meter)

Material: Stainless steel casing with glass measuring tube and synthetic floater.
Specification: Flanged process connection.

Pressure class	Tube diameter in inches	Cost each
Till 300 lbs	1	870.-
	1.5	1,130.-
	2	1,150.-

E

E3 Instrumentation for volume, flow and weight

E3007 Water meter

Material:
- Cast iron housing;
- Internal synthetic mechanism;
- Synthetic accumulator.

Specification: Flanged process connection.

Variation
- Pulse generator + € 100.-.

Diameter in inches	Pressure class in lbs	Cost each
1.5	150	180.-
2	Till 300	280.-
3	Till 300	610.-
4	Till 300	670.-
6	Till 300	1,000.-
8	Till 300	1,300.-
10	150	2,700.-
12	150	3,000.-
16	150	7,200.-
20	150	7,900.-

E3008 Mechanical flow switch

Material: Stainless steel.

Specification:
- Flanged process connection;
- Contact output;
- Adjustable switching point;
- Pressure class up to 300 lbs.

Variation
- EEx d version: + € 210.-.

Connection in inches	Cost each
0.5	920.-
1	990.-
2	1,150.-
3	1,720.-
4	1,720.-
6	2,290.-
8	2,410.-

E3009 Thermal flow switch

Material: Stainless steel measuring element.

Specification:
- Screwed-in process connection;
- Integrated transmitter housing and electronic switches;
- Signal output 4 - 20 mA.

Variation
- Separate installed electronic switches: + € 350.-;
- Flanged process connection: + € 500.-.

Part	Cost each
Switch	1,130.-

E3 Instrumentation for volume, flow and weight

E3010 Electromagnetic flow switch

Material:
- Aluminium housing;
- Stainless steel sensor and connections.

Specification:
- Welding-in socket;
- Contact output;
- Adjustable switching point.

Variation
- Stainless steel housing: + € 500.-.

Part	Pressure class	Cost each
Switch	300 lbs	1,400.-

E3011 Differential pressure transmitter

Material: Stainless steel.

Specification:
- Pressure 150 lbs/120 °C;
- 4 - 20 mA Output with HART protocol.

Excluding:
- Valves block.

Variation
- Flanged model: € 900.-;
- EEx d model: € 120.-;
- Fieldbus: € 230.-;
- Capillary process connection: € 1,000.-;
- Stainless steel case: € 500.-.

Part	Connection	Cost each
Transmitter	1/2'' thread	1,700.-

E3 Instrumentation for volume, flow and weight

E3012 Vortex flow transmitter

Material:
- Stainless steel measuring tube;
- Stainless steel or inconel sensor.

Specification:
- Flanged process connection;
- Signal output 4 - 20 mA with HART protocol;
- Transmitter integrated measuring element.

Variation
- EEx d model: + € 360.-;
- Fieldbus: + € 230.-.

Pressure class in lbs	Tube diameter in inches	Cost each
Till 300	0.5	3,100.-
	1	3,200.-
	1.5	3,400.-
	2	3,500.-
	3	3,700.-
Till 150	4	4,100.-
	6	5,600.-
	8	7,600.-
	10	9,800.-
	12	13,900.-
Till 600	0.5	4,600.-
	1	4,900.-
	1.5	5,100.-
	2	5,200.-
	3	7,900.-
	4	10,200.-
	6	14,500.-

E3013 Variable Area Flowmeter

Material:
- Stainless steel or inconel sensor;
- Stainless steel measuring tube.

Specification:
- Flanged process connections;
- Signal output 4 - 20 mA with HART-protocol.

Variation
- Switch contact: + € 130.-;
- Switch contact for a EEx d model: + € 230.-.

Pressure class in lbs	Tube diameter in inches	Cost each
Till 300	5	1,600.-
	10	1,830.-
	20	2,060.-
	30	2,830.-
	40	3,720.-
Till 600	50	1,730.-
	10	1,890.-
	20	2,170.-
	30	3,280.-
	40	3,950.-

E3 Instrumentation for volume, flow and weight

E3014 Magflow transmitter

Material: Carbon steel.
Specification: Set of two welded neck flanges each with one connecting point.
Variation
- EEx d model: + € 700.-;
- Separate housing for the signal amplifier: + € 500.-;
- Fieldbus: + € 230.-.

Tube diameter in inches	Pressure class in lbs	Cost each
0.5	300	2,900.-
1	300	2,900.-
2	300	3,000.-
3	150	3,100.-
4	150	3,200.-
6	150	4,100.-
8	150	5,300.-
10	150	7,100.-
12	150	9,500.-

E3015 Mass (coriolis) flow transmitter

Material:
- Stainless steel housing;
- Stainless steel measuring tubes (2 pieces);
- Aluminium housing for the signal amplifier.

Specification:
- Flanged process connection;
- Integrated signal amplifier;
- Pressure class 150 lbs.

Variation
- EEX d model: + € 750.-;
- Separate housing for signal amplifier: + € 500.-;
- Fieldbus: + € 350.-;
- Measuring tube of hastelloy (model with single straight pipe) + 50%.

Tube diameter in inches	Cost each	
	1 straight tube	Double tube
0.5	11,100.-	11,400.-
1	11,600.-	11,950.-
1.5	12,600.-	13,100.-
2	13,900.-	14,500.-
3	18,100.-	18,700.-
4		34,300.-

E

E3016 Ultrasonic flow transmitter

Material: Measuring tube of carbon steel.
Specification:
- Clamp model;
- Signal output 4 - 20 mA with HART protocol;
- Pressure class up to 600 lbs.

Variation
- EEx d model: + € 700.-;
- Fieldbus: + € 230.-.

Part in inches	Cost each
1	7,800.-
2	7,900.-
3	7,900.-
4	8,000.-
6	8,100.-
8	10,400.-
10	12,800.-
12	14,400.-

E3017 Thermal flow transmitter

Material: Stainless steel measuring element.
Specification:
- Flanged process connection;
- Integrated transmitter housing;
- Signal connection 4 - 20 mA.

Variation
- EEx d model: + € 320.-;
- Electronic: switch separate installed: + € 330.-;
- HART-protocol: + € 300.-.

Part	Cost each
Thermal flow transmitter	2,800.-

E3018 Positive displacement meter

Material rotating parts: Stainless steel.
Specification:
- Pressure class 150 lbs;
- Measuring of stainless steel or cast iron;
- Internal rotating parts stainless steel;
- Flanged process connection;
- Signal: inductive pulse;
- To calibrate.

Excluding:
- Filter and de-aerator.

Variation
- Counter with round scale included;
- Resettable counter mechanism: + € 1,000.-.

Tube diameter in inches	Cost each	
	Cast iron	Stainless steel
1	3,700.-	6,000.-
2	5,200.-	10,100.-
3	9,700.-	21,100.-
4	12,800.-	32,100.-
6	29,400.-	

E

E3 Instrumentation for volume, flow and weight

E3019 Turbine flow transmitter

Material:
- Instrument housing: Cast iron;
- Internal rotating parts: Stainless steel.

Specification:
- With filter and housing;
- With total counter (gas).

Excluding:
- Pulse generator;
- Hardened stainless steel bearing;
- De-aerator.

Variation
- Impulse amplifier EEx i: + € 240.-;
- Impulse amplifier EEx d:+ € 1,500.-;
- Impulse to mA converter EEx ia/ib: + € 600.-;
- Flow computer mA output: + € 1,000.-.

Part	Pipe diameter in inches	Cost each
Turbine flow transmitter for liquid	1	6,700.-
	2	7,600.-
	3	10,800.-
	4	13,100.-
	6	17,500.-
	8	22,600.-
	10	29,200.-
Turbine flow transmitter for gas	2	3,800.-
	3	4,600.-
	4	4,900.-
	6	7,100.-
	8	8,400.-
	10	14,700.-

E4 Instrumentation for level

E4001 Level glass

Material: Carbon steel.
Specification: Range op to and including 1,200 mm.
Excluding:
- Block-, de-aeration- and drain valves.

Variation
- Safety ball: + € 50.-;
- Lighting (230 Vac, EEx d, 1.274 mm): + € 1,500.-;
- Complete stainless steel: + € 950.-.

Type	Cost each
Transparent	1,900.-
Reflex	2,400.-

E4002 Level glass with magnetic indicator

Material: Stainless steel.
Specification:
- Scale/magnetic lamellae;
- Range 1,000 mm.

Excluding:
- De-aeration- and drain valves.

Variation
- Per 100 mm range extending: + € 25.-;
- Contactors for on/off switching alarm: + € 15.-.

Pressure class in lbs	Diameter in inches	Cost each
150	0.5	980.-
	0.75	980.-
	1	980.-
600	0.5	1,690.-
	0.75	1,690.-
	1	1,690.-

E4003 Floater level indicator

Material: Stainless steel.
Specification:
- Length 300 mm;
- Pressure class lbs 300.

Excluding:
- De-aeration and drain valves.

Variation
- EEx d model: + € 350.-;
- Per 100 mm extra length of probe: + € 150.-.

Diameter in inches	Cost each
1	1,050.-
1.5	1,150.-
2	1,250.-

E4 Instrumentation for level

E4004 Floater level switch

Material:
- Torsion tube: Inconel or monel;
- Floater: Stainless steel.

Specification:
- Range 355 mm;
- Contact outputs;
- Pressure class 300 lbs.

Variation
- EEx d model: + € 350.-;
- Per 100 mm extra length of probe: + € 150.-.

Diameter in inches	Cost each
1	1,500.-
1.5	1,600.-
2	1,700.-

E4005 Vibration fork level switch

Material:
- Vibration fork: Stainless steel;
- Synthetic housing.

Specification:
- Length of probe 118 mm.

Excluding:
- Signal amplifier.

Variation
- EEx d model: € 250.-;
- Per 100 mm extra length of probe: + € 20.-.

Part	Process connection	Cost each
Switch	Screwed G1''	680.-
	Flanged 1'', 300 lbs	940.-

E4006 Capacitive level switch

Material: Stainless steel.

Specification:
- Length of probe 250 mm;
- PTFE Coating on probe;
- Contact output.

Variation
- EEx d model: + € 400.-;
- Flanged connection + € 200.-;
- Profibus + € 230.-.

Part	Cost each
Switch	910.-

E4 Instrumentation for level

E4007 Ultrasonic level switch

Material: Stainless steel.
Specification: Maximum 10 m.
Excluding:
- Signal amplifier.

Variation
- EEx d model: + € 700.-;
- Flanged connection: + € 250.-.

Part	Cost each
Switch	1,690.-

E4008 Radiation based level switch

Material: Stainless steel housing.
Specification: With transmitter/detector.
Including:
- Amplifier (EEx i): € 100.-;
- Power supply (EEx d): € 250.-.

Part	Cost each
Transmitter + holder	6,800.-
Detector + amplifier	6,600.-

E4009 Torsion level switch

Material: Stainless steel.
Specification: Length 100 mm.

Variation
- Extra per 100 mm spindle length: + € 25.-.

Part	Cost each
Switch	360.-

E4010 Differential pressure transmitter

Material: Stainless steel.
Specification:
- 4 - 20 mA with HART-protocol.

Excluding:
- Signal amplifier.

Variation
- Flanged model 3'', 300 lbs: + € 900.-;
- EEx d model: + € 150.-;
- Fieldbus: + € 230.-;
- Process connection with two capillars: + € 1,000.-;
- Case stainless steel: + € 570.-.

Part	Connection	Cost each
Transmitter	1/2'' thread	1,790.-

E4 Instrumentation for level

E4011 Float level transmitter

Material:
- Torsion tube: Stainless steel, Iconel or Monel;
- Float: Stainless steel.

Specification:
- Range 355 mm;
- 4 - 20 mA with HART-protocol;
- Pressure class tot 300 lbs.

Variation
- EEx d model: + € 350.-;
- Per 100 mm range extending: + € 150.-.

Part	Assembly in inches	Cost each
Switch	1	1,670.-
	1.5	1,800.-
	2	1,880.-

E4012 Capacitive level transmitter

Material: Stainless steel.

Specification:
- 4 - 20 mA with HART protocol;
- Isolated probe;
- Screwed process connection;
- Fieldbus: € 230.-.

Part	Cost each
Level transmitter	1,490.-

E

E4013 Ultrasonic level transmitter

Specification:
- Screwed connection;
- Separate ultrasonic transmitter in field case;
- Analogue output and switch output.

Variation
- Fieldbus: + € 430.-.

Measuring distance in m	Cost each
5	1,800.-
9	1,800.-
20	2,300.-
25	2,600.-
45	2,900.-
70	3,300.-

E4014 Radar level transmitter

Material: Stainless steel.
Specification:
- PTFE coating on probe;
- 1.5'' screwed process connection.

Excluding:
- Signal amplifier.

Variation
- 2'' 300 lbs flange connection + € 310.-;
- Fieldbus + € 320.-.

Part	Cost each
Level transmitter free radar	2,200.-
Level transmitter guided radar	2,700.-

E4015 Hydrostatic pressure level transmitter

Material: Stainless steel.
Specification:
- 4 - 20 mA with HART-protocol;
- Screwed process connection.

Variation
- Extra per 100 mm probe length: + € 25.-;
- Fieldbus: + € 200.-;
- Flanged connection: + € 300.-.

Part	Cost each
Level transmitter	1,650.-

E4016 Magnetostrictive level transmitter

Material: Aluminium case, stainless steel sensor tube.
Specification:
- EEx d model;
- 4-20 mA with HART protocol;
- Standard fixed probe;
- Screwed process connection;
- Length 100 mm.

Variation
- Stainless steel case: + € 600.-;
- Flanged process connection: + € 600.-;
- Fieldbus: + € 230.-;
- Sensor tube material Alloy20 or Hastelloy (600 mm): + € 1,900.-;
- Extra per 100 mm length (SS): + € 20.-;
- Extra per 100 mm length (Alloy20/Hastelloy): + € 40.-.

Part	Pressure class in lbs	Cost each
Level transmitter including sensor tube and probe	300 600	3,300.- 3,800.-

E4 Instrumentation for level

E4017 Radiation based level transmitter

Material: Stainless steel.
Specification: Including transmitter/detector.
Including:
- Amplifier (EEx i);
- Power supply unit (EEx d).

Part	Cost each
Transmitter + holder	5,000.-
Measurement tube	7,900.-

E4018 Servo controlled tank measuring system transmitter

Material: Aluminium.
Variation
- High pressure version: + € 250.-.

Including:
- Material Stainless steel;
- Display.

Part	Cost each
Level transmitter, not calibration valence	4,900.-
Level transmitter, calibration valence	10,900.-

E4019 Displacement level transmitter

Material:
- Internal: Stainless steel;
- Case: Cast iron.

Specification:
- Range 100 mm;
- Output 4-20 mA with HART protocol;
- Flanged version (top mounted).

Variation
- EEx d model: + € 250.-;
- Fieldbus: + € 320.-.

Connection in inches	Material	Pressure class in lbs	Cost each
3	Carbon steel	300	2,800.-
		600	3,700.-
	Stainless steel	300	3,200.-
		600	4,100.-
4	Carbon steel	300	2,900.-
		600	3,800.-
	Stainless steel	300	3,500.-
		600	4,300.-

E5 Instrumentation for pressure

E5001 Bourdon pressure gauge

Specification:
- Case diameter up to 100 mm;
- Screwed connection;
- Safety glass.

Variation
- Liquid damping (glycerine): + € 50.-;
- Electric contact device: + € 250.-;
- Chemical seal: + € 250.- to € 750.-;
- Case diameter up to 250 mm (for precision type): + € 500.-;
- Siphon: Stainless steel: + € 180.-.

Type	Specification	Cost each
Bourdon-tube spring gauge	Phosphor bronze Stainless steel	100.- 210.-
Precision tube spring gauge	Phosphor bronze Stainless steel	230.- 340.-

E5002 Pressure switch

Specification:
- Inside adjustable switch point;
- Fixed differential;
- Not explosive resistant.

Variation
- Adjustable differential: + € 50.-;
- Stainless steel measuring element for maximum pressure of 30 bar: + € 60.-;
- EEx d model (max. 30 bar): + € 380.-.

Type	Cost each
Switch	340.-

E

E5003 Differential pressure transmitter with electronic sensor

Material: Aluminium case.
Specification:
- 4 - 20 mA output with HART protocol;
- Screwed process connection.

Variation:
- Flanged model: + € 800.-;
- EEx d model: + € 150.-;
- Fieldbus: + € 250.-;
- Capillar process connection: + € 1,000.-;
- Stainless steel case: + € 600.-.

Accessories:
- 3 way stainless steel valve block with NPT connection: + € 400.-;
- 5 way stainless steel valve block with NPT connection: + € 550.-.

Type	Connection	Cost each
Pressure transmitter	1/2'' thread	1,690.-

E5 Instrumentation for pressure

E5004 Bourdon differential pressure gauge

Specification:
- Case diameter up to 100 mm;
- 2 screwed process connections.

Variation
- Liquid damping (glycerine): + € 50.-;
- Chemical seal: + € 400.- to € 1,000.-.

Type	Specification	Cost each
Bourdon- spring tube pressure gauge	Phosphor bronze	1,320.-
	Stainless steel	1,430.-

E5005 Differential pressure switch

Specification:
- 2 screwed process connections;
- Internal adjustable switching point;
- Fixed differential;
- Not explosive resistant.

Variation:
- Adjustable differential: + € 50.-;
- Outside adjustable switching point: + € 220.-;
- Stainless steel measuring element of max pressure 30 bar: + € 200.-.

Option:
- EEx d model (max. 30 bar): + € 720.-.

Type	Cost each
Switch	1,080.-

E

E5006 Pressure transmitter with electronic sensor

Material: Aluminium case.
Specification:
- 4 - 20mA output with HART protocol;
- Screwed process connection;

Variation:
- Flanged model: + € 400.-;
- EEx d model: + € 150.-;
- Fieldbus: + € 230.-;
- Capillar process connection: + € 800.-;
- Converter for absolute pressure measuring: + € 450.-;
- Stainless steel case: + € 650.-.

Accessoiries:
- 3 way stainless steel valve block with NPT connection: + € 400.-;
- 5 way stainless steel valve block with NPT connection: + € 550.-.

Type	Connection	Cost each
Pressure transmitter	1/2'' thread	1,690.-

E5 Instrumentation for pressure

E5007 Automatic pressure regulator

Variation
- Filter / reducer unit: + € 170.-.

Diameter in inches	Connection in lbs	Material		Cost each
		Housing	Interior	
0.5	150	Bronze	Bronze	940.-
	150	Cast iron	Stainless steel	1,500.-
	300	Cast steel	Stainless steel	1,600.-
1	150	Bronze	Bronze	1,500.-
	150	Bronze	Stainless steel	2,400.-
	150	Cast iron	Stainless steel	1,600.-
	300	Cast steel	Stainless steel	2,500.-
2	150	Bronze	Bronze	2,600.-
	300	Cast steel	Stainless steel	4,400.-
3	300	Cast steel	Stainless steel	2,900.-
4	300	Cast steel	Stainless steel	3,600.-
6	300	Cast steel	Stainless steel	6,700.-

E6 Instrumentation for temperature

E6001 Bi-metal thermometer

Specification: Probe length 200 mm.
Variation
- Probe length 600 mm: + € 100.-;
- Stainless steel model: + € 140.-.

Specification	Cost each
Case, diameter 100 mm sensor of stainless steel	260.-

E6002 Temperature switch with indicator

Specification:
- Electronic contact feature;
- Plug-in unit including fixed fitting 1/2'' screwed connection.

Variation
- EEx d model: + € 300.-.

Description	Cost each
Case, diameter till 100 mm sensor of stainless steel	910.-

E6003 Temperature switch

Variation
- EEx d model: + € 300.-.

Description	Specification	Cost each
Switch with stainless steel sensor	Probe length 200 mm, stainless steel	730.-
	Incl. 2 m flexible capillair, stainless steel	1,340.-

E

E6004 Thermocouple

Specification:
- Stainless steel thermowell, probe length 160 mm;
- Screw-in type.

Variation
- Process connection - flange 1'': + € 250.-;
- With head transmitter built-in, 4 - 20 mA output: + € 300.-.

Description	Cost each
Completely assembled	300.-

E6005 Resistant element (Pt-100)

Specification:
- Stainless steel thermowell, plug-in length 160 mm;
- Screw-in type.

Variation
- Process connection flange 1'': + € 200.-;
- With built in head transmitter, 4 - 20 mA output: + € 300.-.

Description	Cost each
Completely assembled	290.-

E6 Instrumentation for temperature

E6006 Temperature transmitter

Material: Case aluminium.
Specification:
- 4 - 20 mA with HART protocol;
- Electronic integrated in the head;
- With capillary process connection 2 m;
- With 1.5 m flexible, stainless steel capillary tube, measuring element stainless steel.

Variation
- EEx d model: + € 280.-;
- Display unit: + € 300.-;
- Fieldbus: + € 500.-;
- Stainless steel case: + € 300.-.

Type	Specification	Cost each
Transmitter Pt100/thermocouple	Screw thread	750.-

E6007 Thermo wells

Material: Stainless steel.

Variation
- Length 600 mm: + € 80.-;
- Flange connection 1.5'', 150 lbs: + € 150.-.

With the use of the following materials, the associated factors should be used:
- Hastelloy: 4x;
- Iconel: 3.5x;
- Monel: 3x;
- Titanium: 4x.

Type	Specification	Cost each
Tube	Length 170 mm	260.-

E7 Instrumentation for valves

E7001 Globe control valve

Material:
- Cast iron case;
- Internal stainless steel.

Specification:
- 2 way valve;
- Flanged design.

Including:
- Pneumatic actuator;
- Valve position controller 4 - 20 mA.

Variation
- Filter/reducer unit: + € 80.-;
- Internal equipment avoiding cavitation: + 20 up to 100%;
- Internal equipment to reduce sound level: + 20 up to 100%;
- Electrical actuator: see E7007.

Allowances for stainless steel case:
- 1'': + € 1,300.-
- 1.5'': + € 1,800.-
- 2'': + € 2,300.-
- 3'': + € 4,200.-
- 4'': + € 5,900.-
- 6'': + € 12,800.-
- 8'': + € 19,800.-

For additional cost for material and accessories, see chapter B, Piping:
- Extended upper side 2-way and 3-way;
- Handwheels;
- Stellite;
- Materials.

Diameter in inches	Cost each per pressure class	
	150/300 lbs	600 lbs
1	4,400.-	7,600.-
1.5	5,700.-	8,700.-
2	6,500.-	9,800.-
3	8,000.-	12,100.-
4	10,200.-	14,500.-
6	16,200.-	19,200.-
8	20,800.-	25,700.-

E7002 Segment Ball control valves

Material:
- Case cast iron;
- Inside Stainless steel.

Specification:
- 2-way valve;
- Flanged design.

Including:
- Pneumatic actuator;
- Valve position controller 4 - 20 mA.

Variation
- Filter/reducer unit: + € 100.-;
- Inside cavitation protected: + 20 up to 100%;
- Inside to reduce sound level: + 20 up to 100%;
- Electrical actuator: see E7007.

Allowances for stainless steel case:
- 2'': + € 900.-
- 3'': + € 800.-
- 4'': + € 850.-
- 6'': + € 900.-
- 8'': + € 1,950.-
- 10'': + € 2,700.-

For additional cost for materials and accessories, see chapter B, Piping:
- Extended upper side 2-way and 3-way;
- Handwheels;
- Stellite;
- Materials.

Diameter in inches	Cost each per pressure class	
	150/300 lbs	600 lbs
2	4,300.-	6,800.-
3	5,500.-	8,200.-
4	6,400.-	9,500.-
6	10,200.-	13,600.-
8	11,500.-	18,700.-
10	14,900.-	

E

E7 Instrumentation for valves

E7003 Butterfly valves

Material:
- Case cast iron;
- Inside Stainless steel.

Specification:
- Flanged design;
- Pressure class up to 300 lbs.

Including:
- Pneumatic actuator;
- Valve position controller 4 - 20 mA.

Variation
- Electrical limit switch, not explosive resistant: + € 250.-;
- Electrical limit switch, explosive resistant: + € 450.-;
- Filter/reducer unit: + € 100.-;
- Valve position controller 4 - 20 mA: + € 1,100.-;
- Electrical actuator: see E7007.

For additional cost for materials and accessories, see chapter B, Piping:
- Extended upper side 2-way and 3-way;
- Handwheels;
- Stellite;
- Materials.

Diameter in inches	Cost each
3	5,600.-
4	6,500.-
6	7,300.-
8	8,400.-
12	10,800.-
16	17,600.-
20	21,100.-
24	33,800.-

E7 Instrumentation for valves

E7004 Eccentric plug valves

Material:
- Case cast iron;
- Inside Stainless steel.

Specification:
- Flange connection.

Including:
- Pneumatic actuator;
- Valve position controller 4 - 20 mA.

Variation
- Filter/reducer unit: + € 100.-;
- Inside cavitation protected: +20% up to 100%;
- Inside to reduce sound level: +20% up to 100%;
- Electrical actuator: see E7007.

Allowances for stainless steel case:
- 1'': + € 1,600.-
- 1.5'': + € 2,900.-
- 2'': + € 4,400.-
- 3'': + € 5,700.-
- 4'': + € 8,100.-

For additional cost for materials and accessories, see chapter B, Piping:
- Extended upper side 2-way and 3-way;
- Handwheels;
- Stellite;
- Materials.

Diameter in inches	Cost each per pressure class	
	150/300 lbs	600 lbs
1	7,200.-	7,600.-
1.5	7,900.-	7,900.-
2	8,300.-	8,300.-
3	9,200.-	10,100.-
4	10,100.-	11,500.-

E7 Instrumentation for valves

E7005 Ball valves

Material:
- Case cast steel;
- Inside stainless steel.

Specification:
- Full free passage;
- Flange connection.

Including:
- Pneumatic actuator (return stroke by spring).

Variation
- Electrical limit switch, not explosive resistant: + € 200.-;
- Electrical limit switch, explosive resistant: + € 450.-;
- Filter/reducer unit: + € 100.-;
- Electrical actuator: see E7007.

Allowances for stainless steel case:
- 2'': + € 800.-
- 4'': + € 1,200.-
- 6'': + € 1,400.-
- 8'': + € 1,800.-

For additional cost for materials and accessories, see chapter B, Piping:
- Extended upper side 2-way and 3-way;
- Handwheels;
- Stellite;
- Materials.

Diameter in inches	Cost each per pressure class	
	150 lbs	300 lbs
2	3,300.-	3,900.-
4	6,200.-	7,500.-
6	11,700.-	13,600.-
8	16,300.-	18,300.-

E7 Instrumentation for valves

E7006 Membrane valves

Material:
- Case cast iron;
- Inside stainless steel;
- Membrane 300 butyl.

Specification:
- Flange connection;
- Pressure class 150 lbs.

Including:
- Pneumatic actuator.

Variation
- Electrical limit switch: + € 400.-;
- Explosive resistant: + € 400.-;
- Filter/ reducer unit: + € 100.-.

Allowances for stainless steel case:
- 1'': + € 400.-
- 1.5'': + € 600.-
- 2'': + € 700.-
- 3'': + € 1,100.-
- 4'': + € 2,100.-

For additional cost for materials and accessories, see chapter B, Piping:
- Extended upper side 2-way and 3-way;
- Handwheels;
- Stellite;
- Materials.

Diameter in inches	Cost each
1	1,040.-
1.5	1,760.-
2	2,070.-
3	3,210.-
4	4,350.-

E

E7007 Electrical actuators for valves

For prices of valves see B3; Valves and fittings (Chapter B: Piping).
For prices of control- and ball valves see E7001 to E7005.

Diameter in inches	Cost each	
	Loose actuator multiturn	Loose actuator quarterturn
2	3,300.-	4,100.-
3	3,500.-	4,100.-
4	3,700.-	4,100.-
6	4,100.-	5,900.-
8	4,500.-	6,300.-
10	6,500.-	7,000.-
12	6,900.-	7,600.-
14	7,600.-	8,000.-
16	8,100.-	

E8 Installation materials

E8001 Pressure transmitter / indicator direct mounted

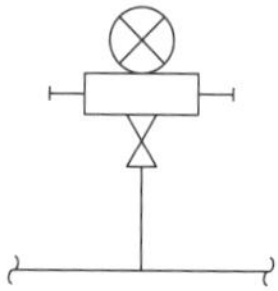

Including:
- Process valve;
- Manifold (2-way) AISI 316L - direct mounted on instrument;
- Installation materials: connectors, instrument cables (from instrument to connecting box and from connection box to junction box), test-connection box, tray, conduit, clamps, connections and name plate;
- Process connecting ring type joint;
- Instrument connection 1/2" NPT.

Excluding:
- Instrument.

Description	Process connection in inches	Cost per hook-up
Hook-up	0.5 1 1.5	1,730.- 2,020.- 2,160.-

E8002 Pressure transmitter/-indicator with 2-way manifold

E

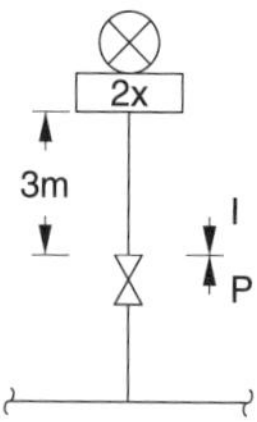

Including:
- Manifold (2-way) AISI 316L;
- 3 m stainless steel pneumatic tube;
- Installation materials: support, connectors, instrument cables (from instrument to connecting box and from connection box to junction box), testconnectionbox, tray, conduit, clamps, connections and name plate;
- Instrument connection 1/2" NPT.

Excluding:
- Instrument.

Variation
- Extra per m pneumatic tube + € 15,-.

Description	Cost per hook-up
Hook-up	1,740.-

E8 Installation materials

E8003 Difference pressure/flow/level transmitter/indicator with 5-way manifold

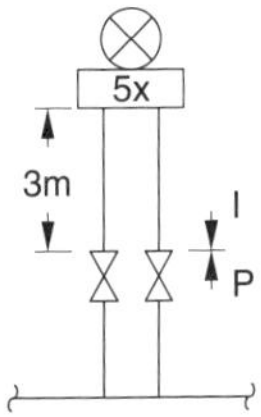

Including:
- Manifold (5-way) AISI 316L;
- 6 m stainless steel pneumatic tube;
- Installation materials: support, installation plate, support connectors, instrument cables (from instrument to connecting box and from connection box to junction box), tes-tconnection box, tray, conduit, clamps, connections and name plate.

Excluding:
- Instrument.

Variation
- Extra per meter length pneumatic tube + € 15.-.

Description	Cost per hook-up
Hook-up	2,060.-

E8004 Control valve (with positioner) and filter/reducer

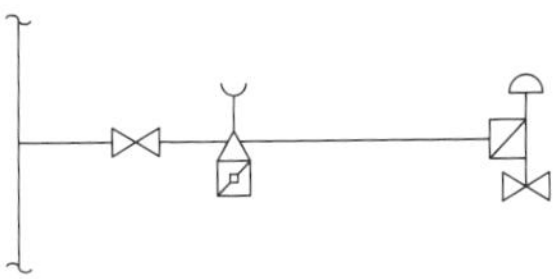

Including:
- Filter/reducer valve;
- 3 m stainless steel pneumatic tube;
- Installation materials: connectors, clamps, mounting brackets, instrument cables (from instrument to connecting box and from connection box to junction box), test-connection box, mounting support, conduit, clamps, connections and name plate.

Excluding:
- Pneumatic controlled valves.

Variation
- Extra per meter length pneumatic tube + € 15.-.

Description	Cost per hook-up
Hook-up	1,810.-

E

E8005 Control valve (with positioner) and filter reducer and magnetic valve

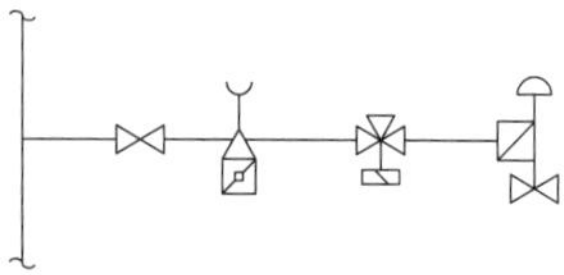

Including:
- Filter/reducer valve;
- Solenoid valve (SS);
- 3 m pneumatic tube (SS);
- Installation materials: installation plate, support connectors, instrument cables (from instrument to connecting box and from connection box to junction box), test-connection box, tray, conduit, clamps, connections and name plate.

Excluding:
- Control valve (see table control valves).

Variation
- Extra per meter length pneumatic tube + € 15.-.

Description	Cost per hook-up
Hook-up	1,870.-

E8006 Valve open/close (with 2 limit switches) and magnetic valve

E

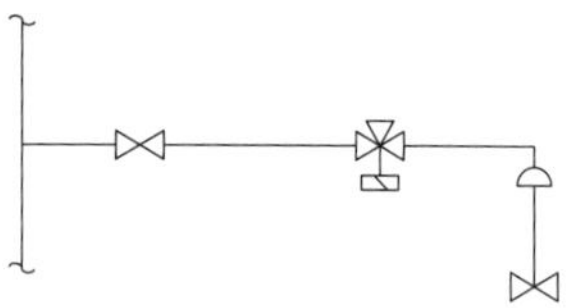

Including:
- Solenoid valve (SS);
- 3 m pneumatic tube (SS);
- Installation materials: connectors, clamps instrument cables (from instrument to connecting box and from connection box to junction box), test-connection box, tray, conduit, clamps, connections and name plate.

Excluding:
- On/off valve.

Variation
- Extra per meter length pneumatic tube + € 15.-.

Description	Cost per hook-up
Hook-up	1,870.-

E8 Installation materials

E8007 Protection cabinets

Material: Polyester.
Specification: Double walled reinforced glass fibre polyester.
Including:
- Doors with safety glass, swivelling handles and standard lock;
- Installation materials.

Variation
- Heating element + € 195.-;
- Explosive resistant heating element + € 720.-;
- Temperature controlled + € 280.-;
- Overpressure safety and security + € 3,200.-;
- Synthetic pedestal + € 770.-.

Dimension in mm (h x b x d)	Cost each
1,050 x 650 x 300	1,770.-
1,050 x 1,050 x 800	2,150.-
1,550 x 650 x 300	2,040.-
1,550 x 1,250 x 800	3,280.-
2,000 x 650 x 300	2,090.-
2,000 x 1,250 x 800	3,510.-

E8008 Junction box

Material: Polyester or aluminium for standard or EEx e boxes.
Excluding:
- Terminals.

Type junction box	Dimension in mm (h x b x d)	Cost each	
		Standard	EEx e
5 - 6 terminals	75 x 80 x 55	40.-	60.-
20 - 26 terminals	75 x 160 x 75	50.-	70.-
20 - 33 terminals	120 x 220 x 90	80.-	130.-
10 - 25 terminals	180 x 180 x 100	110.-	140.-
20 - 30 terminals	230 x 200 x 110	140.-	150.-
35 - 105 terminals	230 x 600 x 110	380.-	440.-

E8009 Instrument cables

Material:
- Blanc electrolytic copper;
- Polyethylene insulation;
- Polyester/aluminium foil shield (common and per couple);
- Outer jacket PVC.

Specification:
- According KEMA/IEC/NEN standards.

Dimension in mm	Cost per m		
	With protection	Without protection	With armouring and protection
1 x 2 x 1.3	6.-		8.-
2 x 2 x 1.3	10.-		12.-
4 x 2 x 1.3	12.-		14.-
8 x 2 x 1.3	16.-		19.-
12 x 2 x 1.3	18.-		24.-
16 x 2 x 1.3	22.-		27.-
24 x 2 x 1.3	31.-		35.-
2 x 1.5		6.-	7.-
4 x 1.5		7.-	7.-
6 x 1.5		9.-	11.-
8 x 1.5		9.-	13.-
12 x 1.5		11.-	15.-
16 x 1.5		13.-	20.-
24 x 1.5		15.-	26.-
30 x 1.5		18.-	31.-
2 x 2.5		7.-	7.-
4 x 2.5		7.-	7.-
6 x 2.5		11.-	12.-
8 x 2.5		13.-	17.-
12 x 2.5		15.-	19.-
24 x 2.5		20.-	37.-
30 x 2.5		27.-	42.-

E

E8010 Price factor Instrumentation cables CPR certified Cca/B2ca

Cost factor for CPR certified cable	
Cable	Cost factor
B2ca	1.2
Cca	1.17

F CIVIL AND BUILDINGS

F0 Explanation

F0001 Explanation of civil and architectural cost data

Mentioned cost data is gross cost data.
The data is including contractor's surcharges and excluding design and engineering costs and value added taxes.
Contractors surcharges are:
- General costs on site;
- General costs;
- Profit and risk.

Cost data can be used only for budget estimating of large projects.
In the various tables, the following abbreviations are used:
- GFA: Gross Floor Area;
- RA: Roof surface area;
- BSA: Building surface area;
- BC: Building Content (volume).

The civil and civil-technical cost data are grouped according to the Dutch civil industry NL/SfB classification system.

The list below shows a review of the codes according to the classification system. For more information about the NL/SfB classification system see the report 'NL/SfB classification system of the Dutch SfB commission' published by the 'Bond van Nederlandse Architecten BNA' in Amsterdam.

NI/SfB code	Description
(0 -)	**Indirect project facilities**
(00.1)	Temporary site facilities
(00.2)	Construction equipment facilities
(00.3)	Risk coverage
(00.4)	Project organization
(00.5)	Home office cost contractor
(00.6)	Temporary facilities
(00.7)	Demolition costs
(1 -)	**Substructure**
(11)	Groundworks
(13)	Floor beds
(16)	Retaining walls, foundations
(17)	Pile foundations
(2 -)	**Primary elements**
(21)	Outer walls
(22)	Inner walls
(23)	Floors, walkways
(24)	Stairs, ramps
(27)	Roofs
(28)	Main load bearing structures
(3 -)	**Secondary elements**
(31)	Outer wall openings
(32)	Inner wall openings
(33)	Floor openings
(34)	Balustrades, railings
(35)	Lowered ceilings
(37)	Roof opening
(38)	System walls
(4 -)	**Finishing**
(41)	Outer wall finishing
(42)	Inner wall finishing
(43)	Floor finishing
(44)	Stairs and ramp finishing
(45)	Ceiling finishing
(47)	Roof finishing
(48)	Final works
(5 -)	**Service, mainly mechanical**
(51)	Heat generation
(52)	Drainage
(53)	Water treatment
(54)	Gas
(55)	Cold generation and distribution
(56)	Heat distribution
(57)	Air treatment
(58)	Climate control

F CIVIL AND BUILDINGS

F0 Explanation

NI/SfB code	Description
(6 -)	**Services, mainly electrical**
(61)	Electrical supply
(62)	Power
(63)	Lighting
(64)	Communication
(65)	Security
(66)	Transport
(67)	Building monitoring
(69)	Structures for installations
(7 -)	**Facilities**
(71)	Traffic facilities
(72)	User facilities
(73)	Food processing facilities
(74)	Sanitary facilities
(75)	Maintenance facilities
(76)	Storage facilities
(8 -)	**Inventory**
(81)	Traffic inventory
(82)	User inventory
(83)	Food processing inventory
(84)	Sanitary inventory
(85)	Maintenance inventory
(86)	Storage inventory
(9 -)	**Site**
(90.1)	Ground facilities
(90.2)	Buildings
(90.3)	Fencing
(90.4)	Site finishing
(90.5)	Site installations mechanical
(90.6)	Site installations electrical
(90.7)	Site facilities
(90.8)	Site inventory

F

F CIVIL AND BUILDINGS

F1 Building complete

F1001 Industrial buildings (manufacturing and workshop buildings)

a. More than 1 floor level and minimum building volume 5,000 m^3.
b. Single floor and minimum height of building 7 m.
Concrete or steel framed construction.
Bare building
Excluding:
- Heating;
- Air conditioning;
- Lighting;
- Pile foundation;
- Asset provisions.

Basic installation
Including:
- Heating;
- Air-conditioning;
- Lighting.

Excluding:
- Pile foundation;
- Asset provisions.

Extensive basic installation
Including:
- Heating;
- Air-conditioning;
- Lighting;
- Asset provisions.

Excluding:
- Pile foundation;
- Furniture.

Model	Cost per m^2 GFA			
	Single floor		Multi level	
	From	Till	From	Till
Bare building	588.-	878.-	888.-	1,280.-
Basic installation	672.-	1,000.-	1,020.-	1,400.-
Extensive basic installation	988.-	1,380.-	1,340.-	1,780.-

F

F CIVIL AND BUILDINGS

F1 Building complete

F1002 Production buildings, additional

Main element	Type	Description	Unit	Cost per unit
(13) Floor beds	Floor duct		m	340.-
(16) Retaining walls, foundations	Concrete middle base		piece	779.-
(22) Inner walls	Fire wall	Lime	m^2	113.-
		Cellular concrete	m^2	124.-
	Fire door		piece	4,600.-
(23) Floors, walkways	Bordes	Concrete, prefab	m^2	165.-
		Steel	m^2	211.-
(31) Outer wall openings	Overhead door	Insulated	m^2	416.-
(37) Roof opening	Sky light	Double-walled	piece	615.-
	Roof light		m^2	237.-
	Smoke hatch	Uninsulated	piece	2,300.-
(59)	Structural facilities	Mechanical	%	4.00
(66) Transport	Lift		piece	21,000.-
(69)	Structural facilities	Electromechanical	%	3.00
(71) Traffic facilities	Deepened loading pit	Not waterproof	positioning place	12,500.-
		Waterproof	positioning place	33,500.-
	Dock plates	Traditional	piece	6,115.-
		Prefab	piece	6,745.-
	Dock leveller		piece	6,115.-
	Dock shelter		piece	2,265.-
	Staircase	Multilayered	floor	7,850.-
(74) Sanitary facilities	Washing and dressing room and toilet group		m^2	1,150.-
(99)	Office		piece	12,000.-
	Boiler house		piece	7,000.-
	Roof construction		m^2	466.-

F

F CIVIL AND BUILDINGS

F1 Building complete

F1003 Laboratory building 3 up to 5 floors

Concrete or steel frame construction.
Bare
Including:
- Elevator;
- Radiator heating;
- Air conditioning/Air exhaust;
- Lighting.

Excluding:
- Pile foundation;
- Furniture.

Furnished
Including:
- Elevator;
- Radiator heating;
- Air conditioning/Air exhaust;
- Lighting;
- Furniture (tables and acid cabins).

Excluding:
- Pile foundation;
- Science laboratory equipment.

Model	Cost per m^2 GFA	
	From	Till
Bare	1,695.-	2,700.-
Furnished	2,230.-	3,200.-

F

F1 Building complete

F1004 Laboratory buildings, additional

Main element	Description	Speciality	Unit	Cost per unit
(24) Stairs, ramps	Main staircase	Simple	floor	9,250.-
		Luxe	floor	13,500.-
	Emergency staircase		floor	5,750.-
(52) Drainage	Sludge pit	5,000 liters	piece	12,500.-
		10,000 liters	piece	18,000.-
(57) Air treatment	Exhaust system		piece	8,000.-
(59)	Architectural provisions for mechanical installations	Mechanical installations	%	6.00
(64) Communication	Telephone exchange	Separate system	connection point	503.-
		Integrated system	connection point	581.-
(65) Security	Electrical sunscreens	Excluding sunscreens	piece	424.-
	Emergency shower		piece	2,425.-
(66) Transport	People/goods elevator		floor	23,000.-
(69)	Architectural provisions for electrical installations	Electrical installations	%	2.00
(74) Sanitary facilities	Toilet group	Simple	piece	20,000.-
		Luxury	piece	27,000.-
	Basement		m^2	924.-
(75) Maintenance facilities	Window cleaning rail (double)		m circumference	322.-
	Window cleaning installation		piece	68,000.-
(78)	Laboratory bench with lower cabinets	Medium	m	1,895.-
		Complicated	m	4,480.-
	Sink with upper cabinets		m	8,695.-
	Acid fume chamber including facilities	Medium	m	7,800.-
		Complicated	m	11,000.-

F1005 Warehouse, free height 7m

Simple warehouse:
- Steel frame construction;
- Tiles on floor;
- Frost resistant heated;
- Simple Lighting;
- Primary fire extinguishing system.

Standard warehouse:
- Steel frame construction;
- Concrete floor;
- Heated;
- Lighting;
- Ventilation.

Excluding:
- Pile foundation;
- Special equipment (sprinkler, fire water collecting system).

Model	Floor	Cost per m^2 GFA	
		From	Till
Simple	Tiles	416.-	505.-
	Stelcon plates (pre-cast reinforced concrete slab)	466.-	565.-
Standard		556.-	740.-

F

F1 Building complete

F1006 Warehouse buildings, additional

Main element	Description	Speciality	Unit	Cost per unit
(16) Retaining walls, foundations	Concrete middle base	Dimensions 2 x 2 x 0.5 m	piece	1,095.-
(22) Inner walls	Firewall	Cellular concrete elements d = 150 mm	m^2	135.-
		Lime stone, d = 150 mm	m^2	126.-
(31) Outer wall openings	External door	Steel, single	piece	1,605.-
	Overhead door	Isolated	m^2	443.-
		Not isolated	m^2	314.-
(32) Inner wall openings	Swing door	Dipterous	piece	3,320.-
	Fire door	Horizontal sliding	piece	4,850.-
(37) Roof opening	Sky light	Single-walled	piece	514.-
		Double-walled	piece	654.-
	Roof light		m^2	254.-
	Smoke hatch	Uninsulated	piece	2,425.-
(59)	Architectural provisions for mechanical	Mechanical installation	%	4.00
(69)	Architectural provisions for electrical	Electrical installation	%	1.00
(71) Traffic facilities	Dock shelter		piece	2,350.-
	Docking board	Traditional	piece	6,400.-
		Prefab	piece	7,000.-
	Dock leveller		piece	6,400.-
	Deepened loading pit	Not waterproof	positioning place	13,000.-
		Waterproof	positioning place	35,000.-
(99)	Toilet group		piece	13,500.-
	Office		piece	12,500.-
	Technical room (boiler house)		piece	7,300.-

F CIVIL AND BUILDINGS

F1 Building complete

F1007 Office building on industrial site, multi level

Office building
Including:
- Elevator;
- Radiator heating;
- Mechanical ventilation;
- Lighting.

Excluding:
- Air conditioning;
- Pile foundation.

Representative office building
Including:
- Elevator;
- Radiator heating;
- Air conditioning;
- Lighting.

Excluding:
- Pile foundation.

Model	Cost per m² GFA	
	From	Till
Office building simple	1,280.-	1,745.-
Representative office building	1,550.-	2,440.-

F1008 Office buildings, additional

Main element	Description	Speciality	Unit	Cost per unit
(84) Sanitary inventory	Toilet group	Simple	piece	20,000.-
		Luxury	piece	27,000.-
(69) Structures for installations	Technical room build on roof-construction	Steelstructure incl. cladding and finishings	m² area technical room	460.-
(24) Stairs, ramps	Main staircase (1 shaft per floor)	Simple	floor	8,950.-
		Luxury	floor	13,500.-
	Emergency staircase (1 shaft per floor)		floor	5,750.-
	Basement	Basic	m²	925.-
	Architectural provisions for mechanical installations	Mechanical installations	%	4.00
(64) Communication	Telephone exchange	Separate system	connection point	503.-
		Integrated system	connection point	581.-
(65) Security	Dry fire extinguisher line		m	166.-
	Electrical sunscreens	Excluding sunscreens	piece	424.-
(66) Transport	Electrical elevators		floor	20,000.-
	Rail (double) for window cleaning installation		m circumference	322.-
	Window cleaning installation	Full automatic	piece	68,000.-
	Architectural provisions for electrical installations	Electrical installations	%	3.00

F1009 Civil and buildings, additional allowances

Description	Allowances in %
Coordination costs for installations	2 - 7
General construction site costs	8 - 15
General costs	7 - 9
Profit and risk	2 - 5

F2 Civil work

F2001 Deepened truck loading pit

Model	Cost per loading area	
	From	Till
Not watertight	18,400.-	23,300.-
Watertight, on piles	45,500.-	49,800.-

F2002 Conveyor bridges

Complete installed.
Excluding:
- Conveyor belt.

Type	Model	Cost per m	
		From	Till
	Open	2,090.-	2,970.-
	Closed	2,660.-	4,650.-
Double	Open	2,660.-	4,440.-
	Closed	3,970.-	6,690.-

F2003 Pipe bridge

Steel, truss pointer pipe bridges.
Span width max. 15 m.
Steel construction (not fire proofed) painted.
Including:
- Pile foundation;
- Provisions for cable trays.

Excluding:
- Treating and dumping cost of contaminated ground;
- Exploratory trench excavation;
- Provisions for underground infrastructure;
- Walkway on pipe bridge;
- Provisions for extension in the future;
- Terrain finishing.

Type	Cost per m	
	From	Till
1-column support with coulings in the longitudinal direction	1,080.-	1,420.-
Small, 1- or 2-layer piping bridge with minor load	2,050.-	2,500.-
Broader, 1- till 2-layer piping bridge with larger horizontal load	3,070.-	3,750.-
Heavier, 3- till 4-layer piping bridge with larger horizontal load	4,080.-	4,770.-

F2004 Parking garage

Capacity of parking places approximately 100 cars.

Type of building	Cost per parking place	
	From	Till
Above ground	14,000.-	18,900.-
Half recessed	17,000.-	24,900.-
Underground	29,000.-	39,900.-

F3 Indirects

F3001 Shed and cabins

Single expendable cost: for installation, furnishing, connecting, dismantling and removal.
Cabin: prefab unit 3 x 8 m.
Shed and Cabin
Including:
- Furnish and furniture.

Part	Unit	One-time cost	Renting cost per week
Shed	m^2	135.-	3.50
Cabin, 3 x 8 m1	each	980.-	125.-
Storage unit/container (100 m^2)	each	775.-	73.-
Toilet unit	each	550.-	105.-
Variation hall	m^2	54.-	1.-

F3002 Scaffolding for building

Including:
- Installation and dismantling;
- Transport and screening inspection.

Type	Unit	Cost per unit	
		From	Till
Facade wall scaffolding	m^2	16.-	24.-
Inner wall scaffolding	m^2	15.-	21.-
Trestle scaffolding	m	8.50	16.-
Rolling scaffold	week	53.-	250.-

F3003 Industrial scaffolding

Scaffolding for industrial work in accordance with DACE Labour Norms.
Remarks
Costs are applicable for 0 - 15 meter and a minimum quantity of 25 m^3.
See H2002: Correction factors for calculations of DACE Labour Norms, chapter General.
Excluding:
- Replacement of disapproved material and lost material;
- Transport cost to and from site.

Including:
- Installation;
- Dismantling.

Activity	Unit	Cost per unit	
		From	Till
Rent	m^3/week	0.40	0.75
Assembly/dismantling standing scaffold	m^3	8.50	
Assembly/dismantling hanging scaffold	m^3	15.50	
Assembly/dismantling extended scaffold	m^3	15.-	
Assembly/dismantling rolling scaffold	m^3	12.-	

F3004 Man hoist and elevators

Including:
- Dismantling and dismantling;
- Transport and screening inspection.

Type	Height in m	Rent		
		Per hour	Per day	Per week
Tower wagon on car	Till 10	62.-	401.-	1,805.-
	15	87.-	561.-	2,540.-
	20	111.-	709.-	3,190.-
	30	165.-	1,065.-	4,845.-
Folding tower wagon (supporting power 500 kg)	Till 7.3		159.-	383.-
	11		270.-	768.-
	15		579.-	1,290.-
Builder's hoist (lifting capacity till 400 kg)	6			105.-
Thread lift (lifting capacity till 1,000 kg)	9			185.-
Rack and pinion lift (lifting capacity till 1,000 kg)	12			490.-
People construction lift (lifting capacity till 1,000 kg)	Till 50			1,200.-

F3005 Mobile cranes

Rent price per hour, minimum rent time variable form 2 - 8 hours, dependent on capacity.
Rent price per week: Week of 40 workable hours.
Including:
- Operation costs;
- Fuel;
- Transport to and from site.

Type/terrain	Boom	Capacity in ton	Cost per hour	Cost per week
Road terrain	Hydraulic	20	105.-	3,670.-
		30	128.-	4,480.-
		35	133.-	4,905.-
		45	152.-	5,710.-
		50	163.-	6,115.-
		80	247.-	9,240.-
		120	342.-	12,930.-
		140	413.-	14,510.-
Rough terrain	Hydraulic	14		3,555.-
		20		3,690.-
		30		4,335.-
		40		4,745.-
Caterpillar	Building	40		3,295.-
		50		4,215.-
		65		5,005.-
		80		6,200.-
		100		7,120.-

F

F3006 Heavy equipment

Rent price per day: Workable day of 8 hours.
Including:
- Operation costs;
- Fuel;
- Transport to and from side.

Part	Specification	Rent per day	
		From	Till
Bulldozer	80 - 270 pk	628.-	992.-
Dragline	Till 1 m^3 bucket capacity More than 1 m^3 bucket capacity	551.- 793.-	683.- 1,190.-
Hydraulic excavator		628.-	926.-
Shovel (loader)	500 - 1,500 liter bucket capacity	528.-	628.-
Fork truck	1 - 7 ton	463.-	662.-
Tractor	50 - 200 pk	396.-	562.-
Truck	Till 10 ton	639.-	717.-
Road roller		661.-	793.-
Vibratory roller		700.-	742.-
Soil compactor	800 - 1,500 kg	424.-	477.-
Compressor including hoses	2.5 - 10 m^3/min	53.-	154.-
Emergency lighting unit	350 kVA	360.-	381.-

F3007 Small equipment

Rent price per week: Week of 40 workable hours.

Part	Rent per week	
	From	Till
Landfill kubel	62.50	83.-
Vibration transducer	198.-	286.-
Vibrator	55.-	75.-
Knock log	50.-	62.50
Measuring instruments	87.-	312.-
Circular saws	55.-	104.-
Industrial vacuum cleaner	87.-	140.-
Hand tools	110.-	218.-
Power tools	150.-	250.-

F3008 Mechanical demolition of complete buildings

Including:
- Transpower of debris.

Excluding:
- Dumping costs;
- Waste treatment;
- Profit on released material.

Type of building	Cost per m^3 BC	
	From	Till
Office and laboratory buildings	13.50	23.-
Simple empty factories and warehouse buildings	8.-	18.-
Heavy production buildings including installation foundations	17.50	35.-

F3009 Mechanical demolition of building elements

Facades: Excluding concrete and steel frame construction.
Roof: Excluding main structure.
Including:
- Transport of debris.

Excluding:
- Dumping costs;
- Waste treatment;
- Profit on released material.

Part	Type of building	Cost per m^2	
		From	Till
Facade	Office- and laboratory buildings	42.-	85.-
	Warehouse buildings, production halls and buildings	26.-	66.-
Roof	Finishing	25.-	70.-

F3 Indirects

F3010 Demolish by hand and wrecking hammer

Including:
- Transport of debris.

Excluding:
- Dumping costs;
- Waste treatment;
- Profit on released material.

Part	Implementation	Unit	Cost per unit	
			From	Till
Concrete structures		m^3	88.-	380.-
Concrete floor	Reinforced	m^3	72.-	360.-
Concrete plinths	Reinforced	m^3	225.-	430.-
	Plain	m^3	88.-	380.-
Masonry	Stretcher bond	m^2	27.-	36.-
	Brick bond	m^2	37.-	83.-
Masonry foundations		m^3	40.-	240.-
Sawing concrete		m	50.-	140.-
Sawing masonry		m	38.50	114.-
Tiled floors and walls		m^2	17.50	54.-
Plasterwork		m^2	9.50	30.-
Partitions		m^2	12.50	33.-
Ceilings		m^2	9.-	16.-

F3011 Dumping costs

Type	Cost per ton	
	From	Till
Sand	8.-	16.-
Ground	8.-	16.-
Lightly contaminated soil	28.-	55.-
Clean debris	7.50	26.-
Coarse debris	9.-	75.-
Rubble with sand	14.-	35.-
Limestone, concrete, gypsum	18.-	69.-
Roofing gravel	28.-	64.-
Roof bitumen	150.-	350.-
Asphalt	8.-	85.-
Wood	36.-	210.-
Insulation	140.-	250.-
Green waste	40.-	205.-
BSA reusable	100.-	210.-
BSA not reusable	195.-	227.-
Industrial	195.-	260.-
Asbestos	110.-	190.-
Chemical waste	160.-	230.-

F4 Building and civil details

F4001 (11) Temporary steel sheet piling / dock plates

Rent including:
- Repair;
- Transport etc.

Method	Price basis	Unit	Cost per unit	
			From	Till
Rent	Purchase minus sale	ton	201.-	366.-
		m^2	31.-	58.-
Piling and removal		m^2	27.-	43.-
Dock plates	Rent per week	m^2	0.70	1.90

F4002 Excavations and backfilling

Activity	Method	Specification	Cost per m^3	
			From	Till
Excavation, loading and removal from site	Mechanically	Excavation > 1,000 m^3 including sampling and landfill cat. 0	13.-	21.30
Excavation and loading	Mechanically	Excavation > 1,000 m^3	6.75	14.25
Excavate	Manually		42.-	72.-
Dispose of surplus soil	Distance depot 3 km	To depot or on site	6.75	14.25
Backfilling	Manually		25.-	37.50
Backfilling and compacting	Mechanically	Exclusive delivery soil/sand	6.50	13.35
Soil supply		From depot on site	7.-	13.-
Sand delivery			18.75	36.10

F4003 Asphalt floors

More than 1,000 m2.
Activities:
- Ground work;
- 300 mm foundation;
- 70 mm gravel asphalt;
- Sealing/tack coat;
- 30 mm bituminous concrete.

Type of floor	Cost per m^2	
	From	Till
Asphalt floors	71.-	104.-

F

F4 Building and civil details

F4004 Concrete floors

Including:
- Work floor;
- Reinforcement;
- Dilatation expansion joints;
- Spread and helicopter surface finish.

Excluding:
- Ground Work;
- Insulation;
- Frost edge, foundation beam;
- Cast in parts.

Type	Specification	Thickness in mm	Cost per m^2	
			From	Till
Concrete floor, not piled	Reinforcement steel (rebar) 60 kg/m^2	150	91.-	101.-
		200	108.-	117.-
		250	134.-	150.-
	Reinforcement steel fibre	200	130.-	150.-
Insulation under floors	Compression-resistant	80	23.-	50.-
	Not compression-resistant	80 - 100	14.-	23.-

F4005 Precast concrete floor slabs (stelcon)

More than 1,000 m^2.

Type of floor	Thickness in mm	Cost per m^2	
		From	Till
Stelcon 2,000 x 2,000 mm	120	71.-	97.-
	140	80.-	105.-
	160	87.-	113.-
Stelcon impermeable, not linked		132.-	164.-
Stelcon impermeable, linked		195.-	235.-

F4006 Work floor (blinding)

Type	Specification	Density in kg/m^3	Cost per m^3	
			From	Till
Compressed concrete	B 15		306.-	411.-
Cellular concrete	Approximately 50 m^3/day	600	117.-	210.-
		900	143.-	227.-
		1,200	158.-	284.-
Expanded Polystyrene Cellular concrete	Approximately 50 m^3/day	350	192.-	323.-

F4007 Foundation constructions

Concrete poured on solid ground.
Including:
- Work floor (blinding);
- Reinforcement.

Excluding:
- Groundwork;
- Insulation;
- Cast in parts, openings, etc.;
- Piles.

Part	Specification	Cost per m^3	
		From	Till
Beams	Diameter < 0.25 m^3	708.-	969.-
	Diameter 0.25 - 0.65 m^3	682.-	831.-
	Diameter > 0.65 m^3	551.-	747.-
Frost edges		551.-	692.-
Footings	Volume < 0.5 m^3	786.-	997.-
	Volume 0.5 - 2.0 m^3	629.-	886.-
	Volume > 2.0 m^3	524.-	775.-
Plinths	Volume < 0.05 m^3	2,350.-	2,600.-
	Volume 0.05 - 0.10 m^3	1,750.-	2,500.-
	Volume 0.10 - 0.20 m^3	1,400.-	2,100.-
	Volume > 0.20 m^3	1,300.-	1,800.-

F4008 Sheet piling, permanent

Material	Part	Weight sheet in kg/m^2	Model	Unit	Cost per unit	
					From	Till
Wood	Sheet piling			m^2	96.-	110.-
Steel	Sheet piling	100 - 150	Exclusive anchor screen	m^2	207.-	287.-
		Above 150	Inclusive anchor screen with anchoring	m^2	374.-	540.-
	Steel purlins, anchors, bolts and welds		Supply and installation	ton	3,300.-	3,750.-

F

F4009 Pile foundations

Minimum number of 50 piles.
Including:
- (Revised) calculations, - drawings, piling plan;
- Measuring and setting out;
- Reinforcement and earthing bars;
- Removal of drilling mud and contaminated ground;
- Acoustic measuring;
- Piles cutting and removal of pile cutoffs.

Excluding:
- Excavation and test trenches;
- Dock plates, fencing etc.

Type	Specification	Cost per m	
		From	Till
Prefabricated concrete piles	Maximal 290 x 290 mm	41.20	55.-
	320 x 320 mm till 380 x 380 mm	45.40	70.-
	Minimal 400 x 400 mm	61.70	90.-
Mortar screw piles	Light construction	85.-	105.-
	Heavy construction	105.-	130.-
Atlas/Fundex/Vibro piles	Light construction	95.-	125.-
	Heavy construction	120.-	150.-
Tubexpiles		250.-	440.-

F4010 Exterior walls and exterior wall finishing

Complete.
Including:
- Inside-outside finishing.

Excluding:
- Exterior wall openings.

Type	Model	Cost per m^2	
		From	Till
Curtain wall		596.-	864.-
Structural glazing		865.-	1,080.-
Brick	2-sided fair facing brick masonry	201.-	269.-
	Cavity wall with insulation	201.-	269.-
Prefab concrete	Washed, with inner leaf	460.-	515.-
Cellular concrete	Painted	147.-	180.-
Steel sheet	Insulated and coated	105.-	145.-
	Uninsulated with steel laths	75.-	90.-
Aluminium plate	Insulated and coated	104.-	141.-
	Uninsulated with steel laths	71.-	83.-
Cavity wall insulation	Mineral wool	22.-	29.-

F4 Building and civil details

F4011 Brick laying or glued interior walls

Including:
- 2 sided flush jointing except glue blocks (clean model);
- Glue blocks shine finishing.

Excluding:
- Inner wall openings;
- Finishing, stucco work and painting;
- Internal scaffolding.

Type of inner wall	Model	Thickness in mm	Finish	Cost per m^2 From	Cost per m^2 Till
Brick	Wall format	100	Facing brickwork	135.-	146.-
			Brickwork	108.-	125.-
		200	Facing brickwork	234.-	264.-
			Brickwork	216.-	249.-
Poriso	Blocks	100	Facing brickwork	95.-	106.-
			Brickwork	89.-	100.-
Lime	Mass format	100	Facing brickwork	115.-	134.-
			Brickwork	89.-	100.-
		200	Facing brickwork	196.-	213.-
			Brickwork	176.-	193.-
	Glue block L	100	Brickwork	84.-	93.-
		150	Brickwork	95.-	113.-
		214	Brickwork	108.-	125.-
Concrete block	Mass format	100	Facing brickwork	173.-	196.-
			Brickwork	107.-	171.-
Gravel concrete block	Blocks masoned	100	Facing brickwork	141.-	158.-
			Brickwork	107.-	145.-
		150	Facing brickwork	167.-	191.-
			Brickwork	126.-	171.-
		200	Facing brickwork	194.-	216.-
			Brickwork	147.-	196.-
Gypsum block	Blocks glued	100	Brickwork	88.-	105.-
Cellular concrete	Blocks glued	100	Brickwork	79.-	153.-
		150	Brickwork	102.-	165.-

F

F4012 System walls

Cost up to 3 m height.
Above 3 m surcharges are applicable.
Including:
- Ready for stucco work, painting and insulation.

Excluding:
- Inside wall openings;
- Finishing.

Type of inner wall	Model	Thickness in mm	Finishing	Cost per m^2 From	Cost per m^2 Till
Metal Stud walls	Plasterboard	100		54.90	74.-
		150		61.80	81.-
		200		84.-	105.-
Metal Stud walls 60 min. fire resistant	Plasterboard	100		92.10	115.-
		150		101.-	124.-
System walls	Light (2-sided plasterboard)		Finished	145.-	180.-
	Heavy (2-sided 2 x plasterboard)		Finished	250.-	295.-

F4013 Concrete slabs supported at both ends

Excluding:
- Floor finishing.

Type	Model	Thickness in mm	Cost per m^2	
			From	Till
Concrete floor	Poured in situation	150	130.-	159.-
		200	147.-	176.-
		250	160.-	192.-
Wide plate floor		150	90.-	121.-
		200	113.-	137.-
		250	130.-	148.-
Hollow core slab	Inclusive compression layer 60 mm	210	90.-	120.-
		260	113.-	135.-
		320	129.-	150.-
Steel plate concrete floor		120	108.-	116.-
		150	120.-	135.-
		180	131.-	146.-
Grate floor	Galvanized steel, high side grid, without supporting construction		145.-	216.-
	Synthetic without supporting construction		129.-	145.-

F4014 Staircases and ladders

Including:
- Finishing.

Excluding:
- Main supporting structure.

Type	Model	Height in m	Unit	Cost per unit	
				From	Till
Prefab concrete straight stairs	b 1,200 mm incl. railing	36	piece	6,500.-	8,000.-
Prefab concrete platform	3 m^2		piece	1,000.-	1,250.-
Steel straight stairs	b 1,000 mm incl. railing	36	piece	6,480.-	7,270.-
Steel spiral staircase	With casing	36	piece	9,930.-	12,500.-
Steel platform	3 m^2		piece	990.-	1,200.-
Climbing ladder	Without cage		m	195.-	210.-
	With cage		m	700.-	800.-
Railing	Post and handrail steel tube		m	300.-	500.-
Wall railing	Steel tube		m	125.-	175.-

F4015 Roof plates and roof decks

Excluding:
- finishing.

Type	Model	Unit	Cost per unit	
			From	Till
Roof plates	Cellular concrete	m^2	71.50	136.-
	Aluminium, kall-zip-system, including edges and drains	m^2	92.-	109.-
	Sendzimir galvanized steel, profiled	m^2	21.-	30.-
Roof decks	Wood (beams + underlayment)	m^2	84.-	96.50

F4016 Main supporting constructions

Part	Type	Model	Unit	Cost per unit	
				From	Till
Steel frame	Production building	Single layer	m^2 GFA	212.-	270.-
		Multi layer	m^2 GFA	133.-	300.-
	Warehouse building	Clear height 7 m	m^2 GFA	128.-	240.-
	Office building	Hollow core slabs	m^2 GFA	111.-	202.-
		Prefab concrete floor plates	m^2 GFA	117.-	225.-
Steel frame		Coated	kg	3.20	4.40
		Galvanized	kg	3.60	5.-
Fire resistant spraying	Beams and columns	60 min. fire resistant	m^2	91.-	142.-
		90 min. fire resistant	m^2	137.-	213.-
Fire resistant coating	Beams and columns	60 min. fire resistant	m^2	60.-	77.-
		90 min. fire resistant	m^2	90.-	129.-
Concrete frame		Concrete poured in situ	m^2 GFA	254.-	460.-
		Prefab	m^2 GFA	255.-	477.-
Concrete wall	Stabilization wall		m^3	960.-	1,350.-
Concrete column	Square		m^3	1,100.-	1,600.-
Concrete beam	Cantilever beam		m^3	1,100.-	1,600.-

F4 Building and civil details

F4017 Doors, windows, facades and awnings

Costs of windows and facades, including insulating glass.

Part	Model	Specification	Unit	Cost per unit	
				From	Till
Windows/facade	Hardwood		m^2	388.-	520.-
	Steel, insulated		m^2	607.-	870.-
	Aluminium, insulated		m^2	505.-	765.-
	Synthetic		m^2	382.-	505.-
Glazing	Insulated	5 - 12 - 4 mm	m^2	84.-	115.-
	Insulated, solar control		m^2	146.-	201.-
Entrance facade	Folding doors	Automatic	piece	12,000.-	17,000.-
Outer door	Steel, insulated	Single, incl. frame	piece	1,450.-	2,150.-
		Double, incl. frame	piece	3,030.-	4,300.-
Entrance facade	Aluminium, incl. glazing		m^2	561.-	805.-
	Sectional doors	Electrically operated	m^2	522.-	570.-
Blinds	Awning	Manual operated	m^2	166.-	250.-
	Vertical cloth	Electrically operated	piece	568.-	760.-
	External Venetian blinds	89 mm lamination	m	230.-	445.-
Windowsill	Stone		m	70.-	135.-
	Artificial stone		m	70.-	80.-
	Tiling		m	30.50	40.-
	Sheet material	MDF	m	38.50	55.-
Curtain cove	Nailed		m	100.-	150.-

F4018 Internal door frames, doors and glazing

Including:
- Painting;
- Hinges and locks.

Part	Model	Specification	Unit	Cost per unit	
				From	Till
Built in	Wooden frame, single	Flush door	piece	677.-	1,050.-
		Lipped door	piece	511.-	540.-
		60 min. fire resisting door	piece	849.-	1,550.-
	Steel frame, single	Flush door	piece	625.-	610.-
		Lipped door	piece	351.-	610.-
		60 min. fire resisting door	piece	897.-	1,525.-
	Assembly frame with rebated door	Anodized aluminium	piece	370.-	640.-
	Steel inner doors including frame	Single	piece	1,475.-	1,750.-
		Double	piece	2,780.-	3,950.-
	Steel fire door	2 x 3 m incl. door closer	piece	4,170.-	4,500.-
	Draught door	3 x 3 m	piece	2,780.-	2,800.-
	Panel wall	Soundproofing	m^2	526.-	670.-
System wall	Without top panel	Flush door	piece	490.-	600.-
		Flush door	piece	744.-	1,100.-
	With top panel	Flush door	piece	585.-	750.-
		Flush door	piece	1,010.-	1,350.-
Internal glazing	Mirrored wired glass	BW30	m^2	144.-	155.-
	Fire resistant glass	Pyroguard C730 BW 30	m^2	430.-	485.-
		Pyroguard C1060 BW 60	m^2	732.-	760.-

F

F4 Building and civil details

F4019 Roof openings

Description	Specification	Unit	Cost per unit	
			From	Till
Sky light	Diameter 1,000 mm	piece	900.-	1,120.-
Roof light		m^2	387.-	870.-

F4020 Outer wall finishes

Description	Model	Cost per m^2	
		From	Till
Painting		13.50	28.-
Plasterwork	Incl. insulation and reinforcement	116.-	157.-
Painting	On stony base	23.-	40.-
Cladding	Natural stone, attached directly	293.-	340.-

F4021 Internal wall finishes

Parts	Model	Cost per m^2	
		From	Till
Jointing		11.-	14.-
Plasterwork	Rough plastering	17.50	20.50
	Float finish	25.-	28.-
	Fine sanding	27.50	43.-
	Spray plaster	10.-	14.-
	Decorative plaster	28.50	52.-
Tile work	Wall tiles	70.-	140.-
Painting	Wall paint on stony base	12.-	14.-
	On wood, coated	29.50	38.-
	On wood, stained	17.50	22.-
	On steel	26.-	34.-
	On concrete (chlorinated rubber paint)	20.50	26.-
Synthetic woven wallpaper	Inclusive painting	26.-	44.-
	Inclusive epoxy paint	34.-	50.-

F CIVIL AND BUILDINGS

F4 Building and civil details

F4022 Floor finishes

Description	Model	Specification	Cost per m²	
			From	Till
Floor covering	Sand cement	30 mm	13.35	19.-
	Monolithic	With sprinkled layer	8.-	14.-
	Mastic asphalt	Red/brown	39.-	56.-
	Asphalt tiles	Thickness 8 mm	37.50	45.-
	Anhydrite	Thickness 50 mm	19.-	25.-
	Magnesite		42.-	46.-
	Hard concrete, D 75	Thickness 50 mm	26.-	43.-
Floor finishes	Epoxy	Not acid resistant	36.-	50.-
		Acid resistant	70.-	110.-
		Mortar floor	65.-	110.-
		Epoxy liquid floor	36.-	50.-
		Epoxy paint	19.-	28.-
	Diced floor	Crosscut	65.-	80.-
	Parquet	Strip parquet	88.-	120.-
		Solid	105.-	190.-
	Linoleum/marmoleum	A-quality	35.50	42.-
		B-quality	29.-	36.-
	PVC		35.-	45.-
	Carpet		33.50	78.-
	Natural stone		125.-	230.-
	Tiles	Double fired tiles	72.50	138.-
	Acid resistant tiles	Till 20 mm thick, grouted with epoxy, layed in cement	140.-	200.-
		On rubber membrane	200.-	285.-
		On membrane in gutters or against plinths	285.-	400.-
		Acid resistant	32.-	55.-
	Rubber		35.-	55.-
	Studded rubber tiles/-band		35.-	55.-
	Oxane oil		9.50	17.50
Making floor dust-free			4.-	8.-
Raised floor system	On leg		145.-	285.-
	On frame		200.-	320.-

F

F4023 Ceiling finishings

System ceiling with visible suspension frame.
Excluding:
- Armatures.

Part	Model	Specification	Cost per m²	
			From	Till
Suspended system ceilings	Mineral fibre	Standard quality	40.-	65.-
		Higher level of finishing	67.50	115.-
	Metal panels	Perforated panels	75.-	130.-
	Metal slats	Open system	57.50	75.-
		Closed system	95.-	133.-
Timbered ceilings	With plasterboard		83.20	105.-
Plasterwork	On plasterboard / stony base		37.15	50.-
Painting	Wall paint		15.-	21.-

F4 Building and civil details

F4024 Roof finishes

Part	Model	Specification	Unit	Cost per unit	
				From	Till
Roofing	Bituminous roofing	1 layer perforated and 1 layer APP-modified	m^2	25.-	36.-
		1 layer perforated and 1 layer APP-modified	m^2	24.-	36.-
	Plastic		m^2	30.-	43.-
	Drain tiles	On carriers	m^2	47.50	51.-
	Low ballast gravel	50 mm	m^2	14.-	19.50
	Low slope of perlite		m^2	30.-	43.-
Roof insulation	PUR foam insulation	PUR 30 till 80 mm	m^2	20.50	37.-
		PS 80 mm till 150 mm	m^2	23.-	38.-
	Mineral wool	100 mm till 150 mm	m^2	37.-	49.-
	Foam glass	90 mm till 130 mm	m^2	60.-	80.-
Edging	Aluminium roof trim	With edging strip	m	34.-	55.-
	Aluminium clamping strip	When working with rising edging strip	m	44.-	90.-

F4025 Sustainable measures, architectural work

Part	Model	Specification	Unit	Cost per unit	
				From	Till
Wall insulation	PIR 190 mm	U=0,13	m^2	48.50	55.-
	PIR 240 mm	U=0,10	m^2	61.-	66.50
Floor insulation	Chape + PUR 170 mm	U=0,13	m^2	37.-	41.50
	Chape + PUR 220 mm	U=0,10	m^2	41.50	46.-
Roof insulation	PIR 170 mm	U=0,13	m^2	63.50	71.-
	PIR 220 mm	U=0,10	m^2	85.50	92.-
	Mineral wool 280 mm	U=0,13	m^2	60.50	66.-
	Mineral wool 330 mm	U=0,10	m^2	73.-	78.50
Window insulation	Double glass	U=1,0	m^2	70.-	94.50
	Triple glass	U=0,6	m^2	116.-	140.-
Sun blinding	Interior screens		m^2	160.-	230.-
	Exterior screens		m^2	280.-	335.-

F5 Building HVAC and Electrical installation

F5001 Heating installations

Description	Model	Unit	Cost per unit	
			From	Till
Warehouse buildings	Heaters	m^2 GFA	40.-	70.-
Office/laboratory buildings, floor height 3,8 m	Radiators	m^3 BC	19.-	27.-
Production buildings	Radiators	m^3 BC	15.-	23.-

F5002 Ventilation and air conditioning

Floor height 3.8 m.
Four-fold ventilation system.

Description	Cost per m^3 BC	
	From	Till
Ventilation	21.-	40.-
Air conditioning	45.-	89.-

F5003 Power installations

From the main low voltage distribution box.

Description	Cost per m^2 GFA	
	From	Till
Light industrial buildings	22.-	28.-
Heavy industrial buildings	42.-	71.-
Warehouses	7.-	11.-
Laboratory	31.-	39.50
Offices	22.-	27.-

F5004 Lighting installations

Including:
- Cable work and armatures.

Excluding:
- Power supply.

Description	Model	Cost per m^2 GFA	
		From	Till
Lighting	250 lux	30.-	45.-
	500 lux	40.-	70.-
	750 lux	65.-	90.-
	1,000 lux	80.-	105.-
Emergency lighting		4.-	6.50

F

F5 Building HVAC and Electrical installation

F5005 Communication installations

Description	Model	Unit	Cost per unit	
			From	Till
Telephone lines	Empty pipes	m^2 GFA	2.50	6.50
Clock installation		m^2 GFA	1	1.-
Sound installation		m^2 GFA	4.-	11.50
Intercom installation		m^2 GFA	2.50	10.50
Cable ducts		m^2 GFA	7.50	12.50
Network and telephone lines	Including cable ducts and hardware	Connecting point	290.-	475.-

F5006 Fire detection and lightning protection

Description	Model	Unit	Cost per unit	
			From	Till
Fire detection	Warehouses	m^2 GFA	12.50	21.-
Fire alert- and evacuation installation	Offices	m^2 GFA	13.50	26.-
Lightning protection		m^2 RA	6.20	17.50

F5007 Fire extinguishing installations

Description	Model	Unit	Cost per unit	
			From	Till
Sprinkler system	Exclusive water buffer stock	m^2 GFA	52.-	84.-
Small extinguishing agent		m^2 GFA	5.-	10.-
Fire hose reel		piece	760.-	960.-

F5 Building HVAC and Electrical installation

F5008 Elevators and window cleaning installations

Transport elevator: Mechanical drive.
Window cleaner ladder: Mechanical drive.
Window cleaner installation: Electrical drive.

Person and goods elevator

Including:
- Doors;
- Frames;
- Cage;
- Rails.

Excluding:
- Shaft;
- Motor;
- Elevator pit.

Transport elevator

Excluding:
- Shaft;
- Pulley room.

Window cleaner installation

Including:
- Rail;
- Elevator cabin.

Description	Model	Specification	Cost each	
			From	Till
Person and goods elevator	4 - 12 stopping places		78,800.-	249,500.-
Transport elevator	1 floor	Lifting capacity 50 - 150 kg	26,000.-	32,000.-
Window cleaner ladder	4 - 8 floors	Maximum 200 m rail	21,000.-	44,000.-
Window cleaner installation		Maximum 200 m rail	64,000.-	160,000.-

F5009 Pantry, wardrobe and desk

Description	Model	Specification	Unit	Cost per unit	
				From	Till
Pantry kitchen unit	Simple	Length 1.5 m	piece	1,000.-	1,100.-
Wardrobe			m	330.-	360.-
Desks	For reception		piece	6,500.-	18,000.-

F5010 Wash basin and water closet combinations

Including:
- Supply and drain pipes.

Description	Model	Cost each	
		From	Till
Closet combination	Floor assembly	595.-	800.-
	Wall assembly	645.-	1,050.-
Wash basin	With mixing tap	525.-	795.-

F5011 Sustainable measures, building services

Part	Model	Specification	Unit	Cost per unit	
				From	Till
Energy production	PV panels	325	Wp/m^2	284.-	354.-
	Heat pump	Cop 3,6	kW	458.-	700.-
		Cop 4,2	kW	502.-	769.-
Free chilling	Heat/cold storage	100 - 200	m^3/h	2,500.-	2,850.-
Combined heat and power	CHP (WKK)	U=0,10	kW	1,000.-	1,300.-

F6 Terrain

F6001 Soil investigation and retaining walls

Description	Model	Specification	Unit	Cost per unit	
				From	Till
Soil investigation	Probing	Till 25 m	piece	189.-	241.-
	Drilling	Till 25 m	m	92.-	166.-
	Adhesion measurement		piece	131.-	143.-
	Setting out and levelling		piece	55.-	74.-
Retaining walls	Prefab, one sided retaining, with coupling strips	1.5 m high, 0.21 m^3 concrete/m	m	342.-	415.-
		2.5 m high, 0.44 m^3 concrete/m	m	503.-	567.-
		3.5 m high, 0.8 m^3 concrete/m	m	751.-	832.-
		5 m high, 1.5 m^3 concrete/m	m	1,245.-	1,330.-

F6002 Environmental soil examination

Description	Specification	Unit	Cost per unit
Soil investigation	NEN 5740/NVN 5740	piece	200.-
Sediment investigation	RIZA package	piece	600.-
Soil drilling	Till 5.0 m-mv	m	59.-
Monitoring well including surround	Supply and installation	m	35.-
Steel protecting cap	Supply and installation	piece	100.-
Sludge sample	Sampling	piece	35.-
Groundwater sample		piece	66.-
pH, Ec, temperature, oxygen, redox potential	Measuring	piece	12.-
Surface samples (excl. hours)		hour	23.-
Sampling container/depot		piece	65.-
Asbestos	Screening soil	piece	120.-
	Screening material NEN 5896	piece	90.-

F6003 Shelter and bicycle shed

Description	Model	Unit	Cost per unit	
			From	Till
Shelter	Length 2.7 m	piece	2,850.-	3,500.-
	Length 5.5 m	piece	5,000.-	6,900.-
Bicycle shed	With paving and lighting	m^2	220.-	240.-

F6 Terrain

F6004 Fencing, gates and barriers

Description	Model	Dimension	Unit	Cost per unit	
				From	Till
Fencing	Tube, wire netting, wire	Height 2 m	m	29.-	43.-
	Bars		m	120.-	189.-
Perkoen timber posts	With three steel wires	Centre to centre 2 m, height 1 m	m	11.50	18.-
Swing gate		Height 2 m, width 1 m	m^2	472.-	632.-
Sliding gate	Single, manual	Height 2 m, width 4 m	m^2	472.-	724.-
	Double, manual		m^2	472.-	770.-
	Electrically operated	Height 2 m, width 9 m	piece	21,000.-	38,500.-
Barrier	Electrically operated	4 m	piece	8,900.-	11,000.-

F6005 Garden work

Description	Model	Unit	Cost per unit	
			From	Till
Landscaping	Including planting, excluding gardening soil and trees	m^2	9.-	18.-
Gardening soil	300 - 400 mm	m^2	17.50	22.80
Lawn	Including milling, harrowing, rolling and sowing	m^2	1.50	4.-

F6006 Surface finishing, hardening

Roads and pavement
Including:
- foundation and sewerage.

Parking places
Including:
- foundation and sewerage system.

Description	Model	Specification	Unit	Cost per unit	
				From	Till
Roads, pavement and hardening	Asphalt	300 kg/m^3 in 3 layers	m^2	88.-	108.-
	Clinker paving	100 - 200 mm sand bedding	m^2	68.-	88.-
	Road construction for heavy traffic		m^2	108.-	131.-
Parking	23 m^2 per parking space	Asphalt	parking space	1,665.-	1,975.-
		Clinker paving	parking space	1,450.-	1,750.-
Tile pavement	Precast concrete pavement slabs	2,000 x 2,000 x 140 mm	m^2	88.-	110.-
		2,000 x 2,000 x 160 mm	m^2	88.-	117.-
	Tiles	With retaining, thickness 40 mm	m^2	36.-	54.-
Pavement	Breaking up, removing and depositing in a heap		m^2	7.-	18.-
Impermeable pavement			m^2	122.-	243.-

F6007 Drainage

Including:
- earth work.

Description	Model	Specification	Unit	Cost per unit	
				From	Till
Flexible tube	Ø 100 - 150 mm	400 - 500 mm deep, backfilling with sands	m	10.60	14.80
		400 - 500 mm deep, backfilling with soil	m	14.-	19.60
Flexible tube with coir	Ø 50 mm	400 - 500 mm depth	m	7.-	16.40
	Ø 100 mm	400 - 500 mm depth	m	11.-	19.60
Control wells	600 x 600 x 500 mm		piece	385.-	425.-
	800 x 800 x 600 mm		piece	500.-	1,400.-

F

F6008 Drainage by well points

Excluding:
- Discharge and draw off costs.

Description	Model	Specification	Drainage depth	Unit	Cost per unit	
					From	Till
Ring installation	Installing and removing	25 insert filters, 100 m drain (ringtube with rapid action coupling)	3 - 6 m	installation	1,695.-	2,500.-
	Using			week	1,700.-	2,600.-
Central pump	Installing and removing			piece	677.-	875.-
	Using			week	465.-	600.-

F6009 Earthwork

Mechanical excavation: Loading and transport over max. 5 km.
Manual excavation: Depending on soil type and depth.
Excluding:
- Dumping fee.

Description	Model	Specification	Unit	Cost per unit	
				From	Till
Excavation	Mechanically	< 1,000 m^3, in layers of 0 - 500 mm	m^3	9.-	15.-
		> 1,000 m^3, in layers of 500 - 1,000 mm	m^3	6.75	14.25
	Manually		m^3	50.-	72.-
Backfilling soil	Manually		m^3	25.-	37.50
Transporting surplus soil	To depot at 3 km		m^3	6.75	14.25
Terrain fill	With sand	< 150 m^3	m^3	29.-	45.-
	With backfill	< 150 m^3	m^3	21.-	30.-

F6 Terrain

F6010 Outside sewer and tanks

Description	Model	Specification	Diameter in mm	Unit	Cost per unit From	Cost per unit Till
Plastic pipes	PVC + attachment, 1 piece / 5 m	Depth 1,000 mm	125	m	28.-	37.-
			200	m	61.-	76.-
			315	m	68.-	84.-
Concrete pipes	Unreinforced, incl. earthwork	Average trench depth 1 m	300	m	80.-	105.-
		No drainage	400	m	90.-	118.-
		Average trench depth 1.5 m	500	m	124.-	159.-
		No drainage	600	m	153.-	196.-
		Average trench depth 2 m	700	m	185.-	227.-
		No drainage	800	m	237.-	278.-
			900	m	290.-	327.-
			1,000	m	342.-	372.-
	Reinforced With welded metal core	Allowance on unreinforced		m	42.-	118.-
		Average trench depth 1 m	300	m	381.-	465.-
			400	m	426.-	521.-
		Average trench depth 1.5 m	500	m	480.-	587.-
			600	m	538.-	657.-
		Average trench depth 2 m	700	m	682.-	833.-
			800	m	705.-	860.-
			900	m	723.-	882.-
		Average trench depth 2.5 m	1,500	m	1,425.-	1,740.-
Poly Ethylene (PE) pipe	Pressureless	Depth 1,000 mm	160	m	14.-	18.-
			200	m	31.-	40.-
			250	m	44.-	56.-
Poly Ethylene (PE) attachments	Pressureless	Pipe bend 90°	160	piece	29.-	34.-
			200	piece	161.-	180.-
			250	piece	373.-	446.-
		Tee 45°	160	piece	60.-	70.-
			200	piece	89.-	111.-
		Electrofusion sleeve	160	piece	24.-	29.-
			200	piece	114.-	131.-
			250	piece	139.-	157.-
		Adapter	200 x 160	piece	137.-	154.-
Glass fibre Reinforced Epoxy (GRE) pipe	Pressure 10 bar, acid resistant	Depth 1,000 mm	80	m	92.-	112.-
			100	m	126.-	142.-
			150	m	141.-	157.-
			200	m	163.-	169.-
Glass fibre Reinforced Epoxy (GRE) attachments	Pressure 10 bar, acid resistant	Bends and flanges, 45°	80	piece	103.-	126.-
			100	piece	124.-	152.-
			150	piece	188.-	229.-
			200	piece	363.-	444.-
		Bends and flanges, 90°	80	piece	116.-	143.-
			100	piece	162.-	198.-
			150	piece	311.-	381.-
			200	piece	436.-	532.-
		Tee, 45°	80	piece	279.-	341.-
			100	piece	323.-	395.-
			150	piece	524.-	640.-
			200	piece	672.-	821.-
		Tee, 90°	80	piece	323.-	395.-
			100	piece	353.-	431.-
			150	piece	588.-	717.-
			200	piece	729.-	889.-
Glue connection of pipes and attachements (GVK - EP)			80	piece	77.-	94.-
			100	verb.	100.-	122.-
			150	verb.	111.-	137.-
			200	verb.	133.-	163.-

F6 Terrain

Description	Model	Specification	Diameter in mm	Unit	Cost per unit	
					From	Till
Manhole	Prefab concrete	Internally 1,000 x 1,000 x 1,000 mm		piece	1,305.-	1,490.-
		Internally 1,500 x 1,500 x 1,500 mm		piece	4,180.-	4,540.-
		Internally 2,000 x 2,000 x 2,000 mm		piece	7,635.-	9,060.-
	Glass fibre reinforced (GVK-EP)	Internally 800 mm, height 1.5 m		piece	958.-	1,170.-
Street/pavement water trap	Concrete / cast iron combination			piece	301.-	368.-
Septic tanks	Prefab concrete	Capacity 5 - 6 m^3		piece	2,060.-	2,340.-

F

G ENGINEERING COSTS

G0 Explanation Estimating engineering cost

G0001 Clarification

DESCRIPTION

In the tables below, the Engineering costs of a project are stated as a percentage of the total costs for the purchase of equipment and material plus the installation (direct and indirect costs of the construction companies). The project costs for the purchase of material and equipment plus installation are set as 100%. The percentages in the table give an indication of a frozen basic design, inclusive follow-up for execution during the construction period, including the as built documentation.

Definitions
Engineering costs are all costs which an engineering office (contractor) makes, and therefore exclude construction management performed by the main contractor and the owner costs. Engineering costs include: man-hours (payroll) for project management and control, engineering and design, procurement and expediting of equipment and materials, the making of construction contracts and all associated costs (non payroll) which the engineering office makes.

For special attention
The ranges are strongly dependent on both the project size and complexity. Generally, as the project value increases, the percentage of the total project cost for engineering will reduce. For small projects, the engineering percentages can vary widely. The ranges given in the tables are for projects from € 500,000.- to € 100,000,000.-. By oil and gas, the ratios are based on carbon steel. By food and pharmaceutical, project engineering is performed by the equipment suppliers. The costs of basic design for onshore transport lines can be higher than for the detailed design.

G

G ENGINEERING COSTS

G1 Estimating engineering costs

G1001 Estimates of engineering costs, (petro)chemical

Type of work	Specifics	Percentage by	
		New	Modification
Manufacturing/process installations	Onshore	12 - 20%	20 - 30%
Underground infrastructure	Distribution lines	8 - 12%	10 - 14%

G1002 Estimates of engineering costs, food/pharmaceutical

Type of work	Percentage by	
	New	Modification
Manufacturing/process installations	7 - 12%	9 - 15%

G1003 Estimates of engineering costs, General Industry

Type of work	Specifics/category	Percentage by	
		New	Modification
Underground infrastructure	Tunnels	12 - 25%	5 - 45%
	Sewerage	10 - 17%	10 - 35%
Above ground infrastructure	Roads	10 - 17%	10 - 35%
	Waterways	5 - 15%	5 - 40%
	Railways	12 - 30%	10 - 50%
Office buildings		7 - 15%	7 - 15%
Industrial buildings		7 - 15%	7 - 15%

G1004 Estimates of engineering costs, utilities/energy

Type of work	Specifics/Category	Percentage by	
		New	Modification
Manufacturing/process installations	Onshore	10 - 15%	12 - 17%
Underground infrastructure	Transport lines onshore	5 - 10%	5 - 10%
	Distribution lines	8 - 15%	10 - 15%

G ENGINEERING COSTS

G1 Estimating engineering costs

G1005 Estimates of engineering costs, oil and gas

Type of work	Specifics/Category	Percentage by	
		Oil and gas	
		New	Modification
Manufacturing/process installations	Onshore	12 - 20%	15 - 25%
	Offshore (exclusive logistics)	18 - 30%	25 - 35%
Underground infrastructure	Transport lines offshore	5 - 10%	5 - 10%
	Transport lines onshore	4 - 8%	4 - 8%
	Distribution lines	8 - 12%	10 - 14%

G1006 Estimates of engineering costs, pulp and paper

Type of work	Percentage by	
	New	Modification
Manufacturing/process installations	8 - 12%	12 - 15%

G

G

H GENERAL DATA

H1 Base material

H1001 Carbon steel

Density 7.85.
Carbon steel 37.2 Din 17100, Fe360 - BEU 25.
Remarks
For orders between 1,000 and 25,000 kg, apply a reduction of € 1.- per 1,000 kg.
For orders up to 75 kg, apply an addition of € 0.20.- per kg.

Description	Quality	Thickness	Cost per 100 kg			
			Range			
			Till 250 kg	From 250 till 500 kg	From 500 till 1,000 kg	More than 1,000 kg
Steel plates up to 3 m wide	S235JR	3 - 10 mm	212.-	182.-	167.-	152.-
		10 - 80 mm	212.-	183.-	168.-	153.-
Chequered floor plate			221.-	191.-	176.-	161.-
Strips			222.-	202.-	182.-	172.-
Hot rolled thin plate	Trade quality	1.5 - 2.75 mm	213.-	183.-	168.-	153.-
Cold rolled thin plate	Trade quality	0.5 - 2.95 mm	220.-	190.-	175.-	160.-

H1002 Carbon steel with surface treatment

Density 7.85.
Carbon steel 37.2 Din 17100, Fe360 - BEU 25. Prices for standard sizes.
Remarks
For orders between 1,000 and 25,000 kg, apply a reduction of € 1.- till € 6.- per 100 kg.
For orders up to 75 kg, apply an addition of € 0.20 per kg.

Description	Quality	Thickness	Cost per 100 kg			
			Range			
			Till 250 kg	From 250 till 500 kg	From 500 till 1,000 kg	More than 1,000 kg
Electrolytically galvanized plates	Trade quality	0.50 - 2.95 mm	239.-	209.-	194.-	179.-
Thermally galvanized flat plates	Trade quality	0.75 - 3.00 mm	233.-	203.-	188.-	173.-
Thermally galvanized corrugated plates	Trade quality		280.-	258.-	233.-	187.-

H GENERAL DATA

H1 Base material

H1003 Profile steel and plates, carbon steel

Density 7.85.
Carbon steel 37.2 DIN 17100, Fe 360 - BEU 25.

Type	Numbers	Cost per 100 kg		
		Ranges		
		From 75 till 150 kg	From 150 till 250 kg	More than 250 kg
Plate-, round-, square-, angle-, T- and small U-steel		208.-	189.-	179.-
IPE	nrs. 80 till 160	183.-	163.-	153.-
	nrs. 180 till 220	182.-	162.-	152.-
	nrs. 240 till 330	185.-	165.-	155.-
	nrs. 360 till 400	187.-	167.-	157.-
	nrs. 450 till 500	195.-	175.-	165.-
	nrs. 550 till 600	202.-	182.-	172.-
UNP	nrs. 80 till 120	183.-	163.-	153.-
	nrs. 140 till 220	183.-	163.-	153.-
	nrs. 240 till 300	184.-	164.-	154.-
	nrs. 320 till 400	199.-	179.-	169.-
INP	nrs. 80 till 220	193.-	171.-	161.-
	nrs. 240 till 500	198.-	173.-	166.-
Broad flange beams HE	nrs. A en B 100 till 180	182.-	162.-	152.-
	nrs. A en B 200 till 220	186.-	166.-	156.-
	nrs. A en B 240 till 300	188.-	168.-	158.-
	nrs. A en B 320 till 600	195.-	175.-	165.-
	nrs. A en B 650 till 1,000	202.-	181.-	169.-
	nrs. M 100 till 1,000	203.-	182.-	169.-

H

H2 Hourly rate

H2001 Hourly rates for installation

Hourly rates consist of the hourly wage plus associated costs which an employer must pay such as social security contributions, holiday surcharges, etc. Also included are allowances for sickness, special leave and turnover of personnel. In addition, there are often other supplements which must also be added. These supplements can vary per company and are dependent on the type of work activities.
The additional allowances are, but are not limited to:
- Costs for use of hand tools;
- Costs for personal protection;
- Costs for consumables, such as gas, oxygen, electricity welding rods, etc.;
- Travel cost between home and work;
- Travel and lodging costs associated with working on site.

When these additional allowances are added, then we have estimated hourly rates for adult and schooled personnel which are applicable for installation activities.
The estimated rates are exclusive:
- Indirect costs such as supervision by working foremen, management costs, overhead, such as company management, insurance, human resources, recruitment and training, IM/IT-support, profit and/or risk;
- Site facilities (site offices, fencing, etc.);
- The use of large equipment such as cranes, heavy transport material etc. (for such items normally separate rentals are used);
- Depreciation costs for machines and equipment (often part of the general costs);
- Access such as scaffolding and mini-lifts;
- Security;
- Warehouse management, etc.;
- Additional uplifts for working in shifts and for hours outside normal day shift.

Remarks
By shift work, the estimated rates per man-hour should have the following % uplift:
- 2-shifts system: 10 - 15%;
- 3-shift system: 15 - 20%;
- 4-shift system: 20 - 25%.

By 4-shift system working on Sundays and public holidays, as well as displaced hours are all included.

Type	Hourly rate	
	From	Till
General construction rate	39.-	55.-

H

H2 Hourly rate

H2002 Correction factor for estimated rates

Certain local conditions, which are beyond the contractor's control, may negatively influence the contractor's efficiency. This makes it necessary to apply correction factors in order to achieve a correct norm time (see DACE Labour Norms) which is representative of the actual conditions in which the work is executed.
Correction factors are applied to the standard norm times to take into account the local circumstances, which are not covered in the standard labour norm.
Often, the costs in the tables within this book are made up of a labour portion (in which a norm-time is estimated) and a material part (where the material costs are estimated).
The correction factor applies only to the labour part. This must be taken into account when applying.
Often in the tables, the labour and material parts are given as percentage of the costs.
The correction factors apply for work activities carried out in such specific conditions.
The local ground level is assumed as 0.

Influence	Description	Circumstances	Factor
Height	Over 6 m	0 to 6 m	1
		6 to 12 m	1.1
		12 to 18 m	1.15
		18 to 24 m	1.2
		> 24 m	1 + 0.01/m
Underground	1.5 m under ground level	0 to -1.5 m	1
		-1.5 to -3 m	1.05
		-3 to -6 m	1.1
Enclosed space		is mainly small, without sunlight and ventilation, and difficult to reach	1.25
Restricted standing height	Creep areas and basements	1.5 to 9 m	1.1
		< 0.9 m	1.2
Special personal protection equipment	If the use of special safety equipment is necessary	Breathing air and acid resistent clothing	2
		Respiratory equipment	1.7
		Safety harnas	1,.25
External circumstances	A location where the influence of the efficiency of labour is know by the owner	For every specific location/space a specific factor shall apply	

H3 Energy

H3001 Tariffs for various energy types

Operating medium	Specification	Heat value	Remarks	Unit	Cost per unit	
					From	Till
Natural gas		31.65 MJ/m^3	More than 150,000 m^3/ year inclusive flat rate	m^3	0.05	0.25
Electricity			From utility company	kWh	0.04	0.15
Oil	Heavy fuel oil Domestic fuel oil	41.03 - 41.55 MJ/kg	Price for 3,500 RWI	GJ GJ	9.- 13.-	19.- 21.-
Propane gas			In bottles of 10.5 kg	bottle	20.-	50.-
Cold water	Drinking water Demineralized water		From utility company Own water supply	m^3 m^3	0.75 750.-	2.- 1,500.-
Steam	High pressure		Exclusive electric power generation	ton	25.-	35.-

H4 SI units

H4001 SI-system of units (NEN 999)

Category	Type of unit	SI-unit		
		Name	Symbol	Derivative
Base units	Length	meter	m	
	Mass	kilogram	kg	
	Time	second	s	
	Electric current	ampere	A	
	Thermodynamic temperature	kelvin	K	
	Amount of substance	mol	mol	
	Luminous intensity	candela	cd	
Supplementary units	Flat angle	radial	rad	
	Solid angle	star radial	sr	
Derived units with own name and symbol	Surface	square meter	m^2	
	Volume, capacity	cubic meter	m^3	
	Frequency	hertz	Hz	s^{-1}
	Force	newton	N	$kg.m.s^{-2}$
	Pressure	pascal	Pa	$N.m^{-2}$ $kg.m^{-1}.s^{-2}$
	Energy	joule	J	$N.m$ $kg.m^2.s^{-2}$
	Power	watt	W	$J.s^{-1}kg.m^2.s^{-3}$
	Electrical conductivity	siemens	S	$V^{-1}.Akg^{-1}.m^{-2}.s^3.A^2$
	Electrical charge	coulomb	C	$A.s$
	Electrical potential difference	volt	V	$W.A^{-1}$ $kg.m^2.s^{-3}.A^{-1}$
	Electrical resistance	ohm	O	$V.A^{-1}$ $kg.m^2.s^{-3}.A^{-2}$
	Electric capacitance	farad	F	$C.V^{-1}$ $kg^{-1}.m^{-2}.A^2$
	Luminous flux	lumen	lm	cd.sr
	Illuminance	lux	lx	$cd.sr.m^{-2}$

H4002 SI-prefixes (NEN 999)

Name of prefix	Symbol	Factor
Yocto	y	10^{-24}
Zepto	z	10^{-21}
Atto	a	10^{-18}
Femto	f	10^{-15}
Pico	p	10^{-12}
Nano	n	10^{-9}
Micro	μ	10^{-6}
Milli	m	10^{-3}
Centi	c	10^{-2}
Deci	d	10^{-1}
Deca	da	10
Hecto	h	10^2
Kilo	k	10^3
Mega	M	10^6
Giga	G	10^9
Tera	T	10^{12}
Peta	P	10^{15}
Exa	E	10^{18}
Zetta	Z	10^{21}
Yotta	Y	10^{24}

H GENERAL DATA

H4 SI units

H4003 Conversions to SI-units (NEN 999 and NEN 3049)

Overview of various SI and imperial units and conversion to SI.

Type of unit	Name of unit	Symbol	Quantity	SI-symbol
Length	micron	µ	10^{-6}	m
	inch	in	0.0254	m
	foot	ft	0.3048	m
	yard	yd	0.9144	m
	mile (statute)	mi	1,609.34	m
	nautical mile	n mile	1,852	m
Surface	centare	ca	1	m^2
	are	a	100	m^2
	hectare	ha	10,000	m^2
	square inch	in^2	645.16×10^{-6}	m^2
	square foot	ft^2	$92{,}903 \times 10^{-6}$	m^2
	acre	acre	4,046.86	m^2
Volume	cubic centimeter	cm^3	10^{-6}	m^3
	cubic decimeter	dm^3	0.001	m^3
	liter	l	0.001	m^3
	register ton	register ton	2.83	m^3
	cubic inch	in^3	16.387×10^{-6}	m^3
	cubic foot	ft^3	$28{,}316.8 \times 10^{-6}$	m^3
	imperial gallon	UKgal	$4{,}546.1 \times 10^{-6}$	m^3
	US gallon	USgal	$3{,}785.4 \times 10^{-6}$	m^3
	dry barrel US	dry bbl (US)	$115{,}627 \times 10^{-6}$	m^3
	oil barrel US	oil bbl (US)	$158{,}987 \times 10^{-6}$	m^3
Flow rate, volumetric flow		$ft^3.h^{-1}$	7.86579×10^{-6}	$m^3.s^{-1}$
		$UKgal.min^{-1}$	75.7682×10^{-6}	$m^3.s^{-1}$
		$UKgal.h^{-1}$	1.2628×10^{-6}	$m^3.s^{-1}$
		USgal.min-1	63.0902×10^{-6}	$m^3.s^{-1}$
		USgal.h-1	1.0515×10^{-6}	$m^3.s^{-1}$
		oil bbl.d-1 (US)	1.84013×10^{-6}	$m^3.s^{-1}$
Mass	gram	g	0.001	kg
	grain	gr	64.7989×10^{-6}	kg
	pound	lb	$453{,}592 \times 10^{-6}$	kg
	ton	t	1,000	kg
	UK ton	UKton	1,016.05	kg
Density	kilogram per cubical meter			$kg.m^3$
		$lb.ft^{-3}$	16.0185	$kg.m^{-3}$
		$lb.UKgal^{-1}$	99.7763	$kg.m^{-3}$
		$lb.USgal^{-1}$	119.826	$kg.m^{-3}$
Force	kilogram force	kgf	9.80665	N
	poundforce	lbf	4.44822	N
	poundal	pdl	0.13826	N
Moment of force	newton meter			N.m

H4004 Conversions to SI-units (NEN 999 and NEN 3049), continued

Overview of various SI and Imperial units and conversion to SI.

Type of unit	Name of unit	Symbol	Quantity	SI-symbol
Pressure	atmosphere (normal)	atm	101,325	Pa
	technical atmosphere	at	98,066.5	Pa
		$kgf.cm^2$	98,066.5	Pa
	Pound-force per square inch	Psi	6,894.76	Pa
	bar	bar	100,000	Pa
	mm Mercury column	mm Hg	133.322	Pa
	inch Hg (mercury)	in Hg	3,386.39	Pa
	mm water column	mm H_2O	9.80665	Pa
	inch water	in H_2O	249.089	Pa
	foot water	ft H_2O	2,989.07	Pa
	torr	torr	133.322	Pa
Work, energy, amount of heat	calorie	cal	4.186	J
	newton meter	N.m	1	J
	watt second	W.s	1	J
	kilowatt-hour	kWh	3,600,000	J
	British thermal unit	Btu	1,055.06	J
	horsepower hour	hph	2,685,128	J
Power, energy flow, convective heat	joule per second	$J.s^{-1}$	1	W
		$kcal.h^{-1}$	1.163	W
	horsepower	pk	735.499	W
	British horsepower	Bhp	745.7	W
		$Btu.h^{-1}$	0.29307	W
Electrical power				
- Effective	watt	W		W
- Apparent	volt.ampere VA	VA	1	W
- Reactive (blind capacity)	var	var	1	W
Specific heat	joule per kilogram per Kelvin			$J.(kg.K)^{-1}$
		$kcal.(kg.C)^{-1}$	4,186.8	$J.(kg.K)^{-1}$
		$Btu.(lb.F)^{-1}$	4,186.8	$J.(kg.K)^{-1}$
Heat load		$kcal.(m^2.h)^{-1}$	1.163	$W.m^2$
		$Btu.(ft^2.h)^{-1}$	3.15459	$W.m^2$
Heat conductivity coefficient		$kcal.(m.h.C)^{-1}$	1.163	$W.(m.K)^{-1}$
		$Btu.(ft.h.F)^{-1}$	1.73073	$W.(m.K)^{-1}$
Heat transfer coefficient		$kcal.(m^2.h.C)^{-1}$	1.163	$W(m^2.K)^{-1}$
		$Btu.(ft^2.h.F)^{-1}$	5.67826	$W(m^2.K)^{-1}$

H

H4005 Conversions to SI-units (NEN 999 and NEN 3049), continued

Overview of various SI-units and imperial units and conversion to SI-units.

Type of unit	Name of unit	Symbol	Quantity	SI-symbol
Angular velocity	radial per second			$rad.s^{-1}$
Velocity	meter per second			$m.s^{-1}$
	foot per second	$ft.s^{-1}$	0.3048	$m.s^{-1}$
	mile per hour	$mile.h^{-1}$	0.44704	$m.s^{-1}$
Acceleration	meter per square second			$m.s^{-2}$
	foot per square second	$ft.s^{-2}$	0.3048	$m.s^{-2}$
Time	minute	min	60	s
	hour	h	3,600	s
	day	d	86,400	s
Temperature (difference)	degrees Celsius	C	1	K
	degrees Fahrenheit	F	0.55556	K
Temperature (absolute)	0 degrees Celsius	C	273.15	K
	0 degrees Fahrenheit	F	255.372	K
(Flat) angle	degrees	°	$17{,}453.3 \times 10^{-6}$	rad
Quantity of light	lumen second			lm.s
Illumination strength	lumen per square meter	$lm.m^{-2}$	1	lx
	lumen per square foot	$lm.ft^{-2}$	10.7639	lx
	foot candle	fc	10.7639	lx
Illumination	lux second			lx.s
Luminance	candela per square meter			$cd.m^{-2}$
Viscosity (dynamic)	Pascalsecond			Pa.s
		$kgf.s.m^{-2}$	9.80665	Pa.s
		$kg.(m.s)^{-1}$	1	Pa.s
		$lb.(ft.s)^{-1}$	1.4882	Pa.s
		$lb.(ft.h)^{-1}$	413.379×10^{-6}	Pa.s
	centipoise	cP	0.001	Pa.s
	poise	P	0.1	Pa.s
Viscosity (kinematic)	stoke	St	0.0001	$m^2.s^{-1}$
		$ft^2.s^{-1}$	0.0929	$m^2.s^{-1}$

H5 Index figures

H5001 General

GENERAL

The indices must be seen as an indication of the change of the actual prices and are based on domestic consumption (domestically produced and imported products) unless specifically otherwise stated.

The indices in this chapter have been collected by DACE Special Interest Group Cost Engineering Process Industry (CEPI, previously known as WEBCI and WUBO).

The series in table H5003 are based on prices excluding taxes and are taken from the CBS (Statistics Netherlands), in accordance with the goods classification PRODCOM. The series 30 and 32 are based on the MBK-index for the construction industry.

All indices have a base year 2015 = 100. Indices for 2021 are provisional figures, determined by the CEPI based on CBS figures to and including September 2021.

When applicable, the figures have been recalculated using 2015 as the base year. The use of a new base year is often announced too late, sometimes even years later. During the intervening periods the CBS uses the old figures. This is taken into account in the published indices.

Series 60 is put together by the CEPI for process installations in the Netherlands. The series is - as far as the material component is concerned - based on CBS calculated producer price index figures. For the wage component the CPB (Netherlands Bureau for Economic Policy Analysis) published data on wages and labour productivity development have been used. Some of this data has been adjusted retroactively where necessary. The weighing factors have been based on the WEBCI 1985 research.

The non-Dutch series in table H5002 are based on prices in national currencies and are derived from 'Chemie Technik' (Germany), 'Chemical Engineering' and the US Bureau of Labor Statistics (USA). The results of the effects of taxation on the figures are not available. All series have been recalculated using the basis year 2015 = 100.

Series 70 and 75 apply for apparatus and machines for chemical and related associated process installations and are exclusive installation.

Series 85 to 91 are applicable for process installations and are inclusive design and installation costs. For series 85 (process installation Germany) the build up in 2010 has changed resulting in minimal differences in the figures.

H5 Index figures

H5002 Index figures outside the Netherlands from 2015

Base year 2015 = 100.

Item	Series	2015	2016	2017	2018	2019	2020	2021
70	Apparatus and machines Germany	100	100	104	103	105	104	106
75	Apparatus and machines United States	100	101	101	103	101	103	104
85	Process installations Germany	100	102	105	107	105	104	106
90	Process installations United Status chemical industry	100	94	105	107	105	105	109
91	Process installations United States oil refineries	100	97	96	100	104	80	96

H5003 Index figures from the Netherlands from 2015

Base year 2015 = 100.

Item	Series	2015	2016	2017	2018	2019	2020	2021
1	Water (domestic)	100	97	96	90	91	93	93
2	Natural gas	100	101	103	111	124	126	126
3	Fuel oil	100	94	100	105	107	78	78
4	Electricity	100	85	86	100	116	70	70
10	Steel plates	100	-	-	-	122	106	106
12	Steel beams	100	-	-	-	-	-	-
13	Steel pipes	100	98	108	115	114	110	110
20	Wages per worked hour (metal-electrical)	100	102	104	106	108	107	107
25	Consumer consumption, all households	100	100	102	103	106	108	109
30	Buildings (office buildings)	100	99	98	102	106	109	109
31	Steel constructions	100	101	107	109	111	112	112
32	Building materials, freight paid	100	-	-	-	-	-	-
41	Machines	100	101	101	102	104	106	106
44	Motors and turbines	100	100	100	103	105	105	105
45	Pumps and compressors	100	103	101	101	101	104	104
46	Tanks and reservoirs	100	101	101	103	104	107	107
48	Accessories	100	-	-	-	-	-	-
50	Electric motors and generators	100	-	99	100	101	101	101
51	Electric motors (import)	100	-	-	-	-	-	-
52	Transformers	100	-	95	96	97	98	98
53	Steam boilers	100	98	99	99	100	102	102
55	Hoist-, lift- and transport works	100	102	103	105	108	108	108
60	Webci composite index	100	-	-	-	-	-	-